D0065013

Cellular Radio Systems

The Artech House Mobile Communications Series

John Walker, *Series Editor*

For a complete listing of *The Artech House Telecommunications Library,*
turn to the back of this book

Cellular Radio Systems

D. M. Balston and R. C. V. Macario

Editors

Artech House
Boston • London

Library of Congress Cataloging-in-Publication Data
Balston, D. M.
Cellular Radio Systems/D. M. Balston and R. C. V. Macario, editors
Includes bibliographical references and index.
ISBN 0-89006-646-9
1. Cellular radio. I. Balston, D. M. II. Macario, R. C. V. (Raymond Charles Vincent)
TK6570.M6C47   1993   93-31141
621.3845'6–dc20   CIP

© 1993 ARTECH HOUSE, INC.
685 Canton Street
Norwood, MA 02062

All rights reserved. Printed and bound in the United States of America. No part of this book may be reproduced or utilized in any form or by any means, electronic or mechanical, including photocopying, recording, or by any information storage and retrieval system, without permission in writing from the publisher.

International Standard Book Number: 0-89006-646-9
Library of Congress Catalog Card Number: 93-31141

10 9 8 7 6 5 4 3 2 1

Table of Contents

Part I
Cellular Radio Principles

Chapter 1
Cellular Radio Principles

R. C. V. Macario
University College of Swansea

1.1 OBJECTIVES OF CELLULAR RADIO

Cellular radio has come about because of the need to provide mobile telephone service nationally, not just statewide. The conventional public telephone service, usually known as the *public switched telephone network* (PSTN), provides national and international coverage by means of a fixed network switching and numbering hierarchy. Decoupling the telephone from its wires to the local exchange (LE) to provide a radiotelephone using radio communication channels has been explored for several decades.

The fundamental problems with such a concept are radio range, or coverage, and number of channels, or voice circuits. These in turn conflict with two aspects of any conceptual national radiotelephone service: full, seamless service coverage and the likelihood of a large number of subscribers.

The word *subscriber* arises because the user of the radiotelephone or mobile telephone needs to subscribe to the cost of the infrastructure, which is a major part of providing the service whether the user has purchased the radiotelephone instrument or not. Subscribers to cellular radio can be counted in millions, so we are clearly talking about millions of radio channels within systems that simultaneously provide national coverage.

Several fundamental attributes are needed for the realization of this service:

1. Frequency agility in the radiotelephone;
2. A contiguous arrangement of radio cells so that the mobile units can always operate at acceptable radio signal levels;
3. The roaming feature by which mobile units can have continuous service as they roam the geographical area of the service—an area that is usually national;

4. A fully integrated transparent fixed network managing these operations.

These matters are explained in general in this introductory chapter, but full and particular system details will be found in later chapters as well as in [1–11].

First of all, however, in every situation there is the *mobile station* (MS), which denotes the radiotelephone, (or *handset* as the mobile station is often called), and the *base station* (BS) with which radio connection takes place both ways. Because there will be many mobiles and base-station centers covering the area of service that most likely will be operating at different radio frequencies (channels), it is essential that the mobile unit can operate on any of the channels as commanded by its base station; that is, mobiles must have the property of *frequency agility*. (New *spread-spectrum systems* may not appear to need this frequency agility, but the sets are, of course, continuously scanning a wide range of the spectrum.)

Once the mobile part has this frequency ability (item 1 of the fundamental attributes), two of the fundamental problems listed above can be solved. First, the subscriber's service continues if he or she moves out of the range of one base station area because the subscriber equipment can switch to a new channel operating from a new site, and, for all intents and purposes, there is seamless, continuous coverage wherever the user roams. Second, other subscribers can use other channels throughout the state or country, and this has a multiplying effect so that the system appears to have a larger number of radio channels. The effect is known as *trunking gain* and allows a much larger subscriber base than at first might be expected.

This performance is not sufficient on its own, however. All four attributes are equally necessary. For example, regarding item 2, the layout of the radio cells requires continuous but noninterfering coverage: another paramount feature of cellular radio. In *frequency reuse*, clusters of radio cells, usually shown as regular hexagons, manage the allocated radio frequencies assigned to the cellular radio service in relation to customer needs. The (item 3) roaming feature can then be automatically introduced because customers will no longer run out of radio spectrum as they roam through individual cells and into new cell clusters, since there are continuously allocated alternative duplex voice (radio) channels.

This whole strategy quite clearly needs managing, and this is where the *fixed network* of cellular radio comes in. Base stations as well as cell clusters must be linked together. Registration centers usually keep track of the ''who, where, and which'' of subscribers. At the same time, the fixed network must do the call routing within, into, or out of the network. Like the PSTN, a cellular radio network has a complex switching and numbering hierarchy.

These cellular radio principles can be seen within all the varied systems, whether analog, digital, or hybrid. Each national or regional system has a special history as well as special operational features. The following chapters describe these massive enterprises in a structured way. First, however, a discussion of topics fundamental to cellular radio is offered.

1.2 FUNDAMENTAL FEATURES

1.2.1 Mobile Station

The radiotelephone is defined as a telephone without wires; the connection to the local exchange is made via radio. To achieve operation, many new factors above and beyond the basic telephone need to be appreciated. Figure 1.1 shows, in a semidescriptive way, the features of the basic telephone. It has a number, which is registered solely in the local exchange. Numbers can be dialed from the phone using a keypad and the accepted dual-tone multifrequency (DTMF) format. Power to operate the ringer and other functions in the phone is supplied by the local exchange through the local loop. Usually, only two wires are used; thus, the length from the local loop to the exchange is limited because amplification is not possible with only two wires and two-way conversation (duplex) is expected (another fundamental feature of a telephone).

Another feature of the basic telephone is the cradle, which alerts the local exchange as to whether the phone is *on-hook* or *off-hook*. To make the telephone system work, many other features, of course, are to be found and are necessary within the fixed telephone network system part.

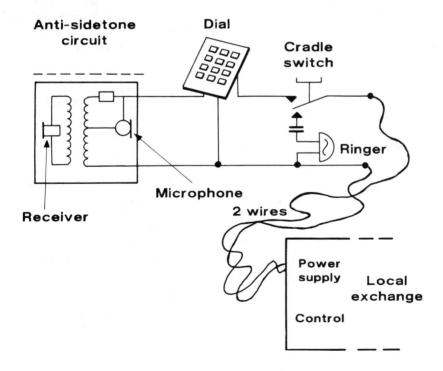

Figure 1.1 The basic telephone with a two-wire connection to the local exchange.

The radiotelephone is quite different, apart from the microphone and earpiece. Again, a semidescriptive diagram is a good place to start. With reference to Figure 1.2, a radiotelephone clearly differs in many ways. For example, it needs a rechargeable battery to supply power for the operation of the unit. The local exchange is replaced by a local base station. Now the base station is connected to the mobile telephone network—similar to the fixed "part" mentioned earlier.

Both the mobile station and the base station need radio antennas. These antennas must be suitable for the radio frequencies that are licensed and allocated within the radio spectrum set aside for the service. Generally, two radio channels must be allocated to each mobile for duplex operation; that is, the user expects to speak and listen at the same time. A variation on this configuration that uses time-division multiplex operation is described later; but, whatever the case, forward and return path radio channels are required. The *forward channel* refers to the base-to-mobile path, and the *reverse channel* refers to the mobile-to-base path. Both the base station and the mobile station require radio transmitter circuits. The weakest path is the reverse (or return) path because it is necessary to save on transmitter power and hence battery recharge cycle time in the mobile station.

The mobile station now carries its own "telephone" number in internal memories. This is unique to a personal radiotelephone or cellular radio. It allows the subscriber to roam over the network—provided an agreement can be set up with the local base station where the subscriber is operating. (This roaming feature was briefly described earlier.) Put another way, provided a mobile radiotelephone network and service has been set up

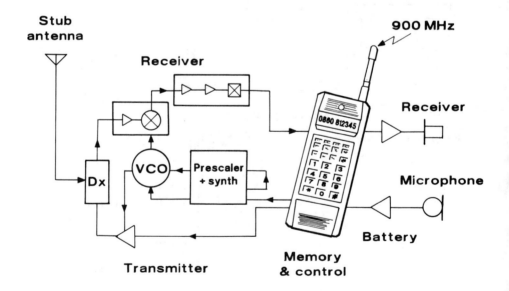

Figure 1.2 Outline of the components within a cellular radiotelephone or mobile station.

by a network operator and the holder of the radio license, the subscriber (to the service, the user of the mobile station) can roam around the network or the country as long as he or she stays within radio distance of a base station.

The mobile station will also contain a radio receiver, as well as a transmitter and tuning (synthesizer) circuits, supplied with power from the battery, and will take and give instructions to the local memory and the control module. The "ringer" is now replaced and controlled through the above circuits. No cradle is found on a radiotelephone; the base station and the mobile station automatically keep in touch with various handshaking signals. A call is set up by depressing a specific SEND button on the keyboard, which is not associated with the DTMF keyboard instructions.

It is clear that a radio or cellular telephone handset is completely different technologically from a fixed telephone instrument, and not just because the wire to the local exchange is missing. The sets also differ in a large number of operational features.

Our subscriber has a telephone number, now called a mobile subscriber number, which is registered to the handset. The number will usually be the same format as numbers allocated to fixed subscribers, except that the area or office code will not refer, for example, to a specific town or district, but to a specific mobile telephone service.

The fixed network will hold a register of mobile numbers for billing, authentication, and location purposes, but the mobile phone number specific to the handset does, of course, go hand in hand with the roaming feature.

Finally, for call management, much more sophisticated signaling between the mobile and local base stations (and supporting network) is necessary, and specific instructions must be given by the fixed network part, through the base station. Thus, a large amount of digital data traffic is found in cellular radio whether the system is analog or digital.

1.2.2 Cellular Radio Cell

A single radio cell and the factors that dictate coverage are illustrated in Figure 1.3. The base station will usually be placed in a clear, commanding position and will have an appropriate transmitter power (say, in excess of 10W), a sensitive receiver with a low noise figure, minimal site noise, and useful antenna gain. The mobile station will have limited transmitter power, especially in the portable mode, and an elementary antenna. The base station receiver performance is unable to make up this 20-dB or so loss of return path signal power, and it is therefore the reverse path that limits the radio range.

Three ranges are indicated in the diagram.

1. The operating range: distance d;
2. The maximum radio range (i.e., cell size limited by noise, propagation factors, and transmitter power): R_{max};
3. The cell radius designed for the system, which will be less than R_{max} (decided by the coverage and cell pattern considerations, described below)—called R.

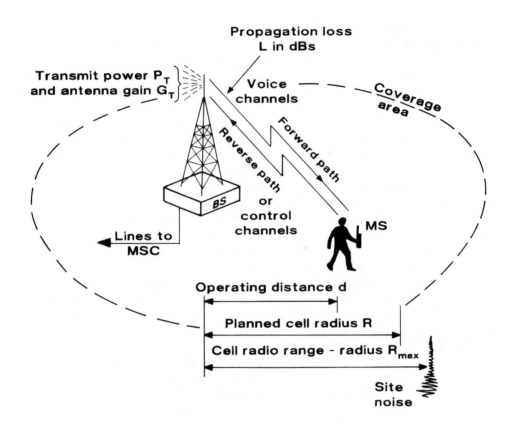

Figure 1.3 The fundamental radio cell and associated parameters showing a base station and mobile station.

For a flat terrain, R can be regarded as the radius of a circle making up the cell. The area covered will thus be πR^2. For example, a cell radius of 1 km will cover 3.14 km²; a cell radius of 10 km will cover 314 km² (a small town).

In practice, the cell coverage area is not circular. It will often be distinctly misshapen; thus, the parameter R does not necessarily apply. Figure 1.4 shows two examples: (a) is for an uneven terrain or terrain with distinct obstacles, such as hills or buildings; (b) is when a directional antenna is used at the base station.

The directional antenna arrangement allows either (1) cell *sectorization* or (2) *corner* illumination. Examples of these strategies appear later in the text.

It would, however, be very difficult to plan continuous national coverage using these irregular cell coverage shapes. For example, a group of circular cells overlap quite extensively if no area is to be without coverage (see Figure 1.5). This type of arrangement is nevertheless important for airborne radiotelephone services.

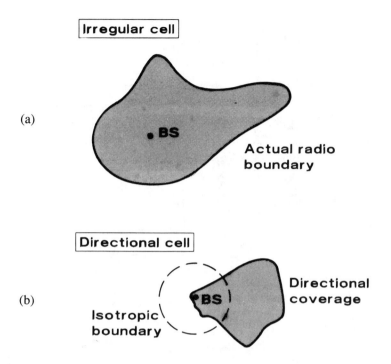

Figure 1.4 Practical cell coverage concepts: (a) uneven terrain in the vicinity of base station; (b) base station has a directional antenna.

1.2.3 Multiple Cell Layout Geometry

Operation in a single cell is a well-understood aspect of private mobile radio (PMR). The use of a repeater (transmitter/receiver) can be likened to a central, well-placed base station, so large areas can be covered by a repeater. This is especially important for rural areas, but for a low-power public radiotelephone concept of service, cell sizes will be smaller, and it is important to note that the same frequencies cannot be repeated in adjacent cells.

A particularly characteristic feature of cellular radio is the association of the hexagonal (six-sided polygon) with each radio cell surrounding a base station. Like triangles and squares, hexagons can be tessellated, that is, laid out side-by-side, giving seamless coverage. A hexagon is not that different from a circle, as can be seen in Figure 1.6. Also, the concept of sectorization is made almost natural by viewing the 120-deg angle, equal to 3 sectors, or 60-deg angle, equal to 6 sectors. A directional antenna placed at the base station site (Figure 1.3) will achieve this sectorization.

Returning to uniformly illuminated cells, however, Figure 1.7 shows how hexagonal cells can be laid out in clusters, which then can be repeated to cover larger areas. Examples of cluster sizes 3, 4, and 7 are shown in the top half of the diagram.

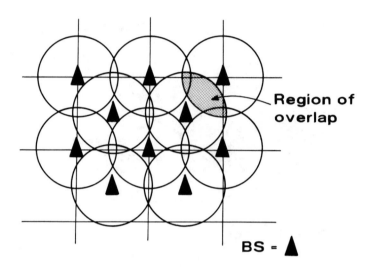

Figure 1.5 Radio cell plan set out on a square grid.

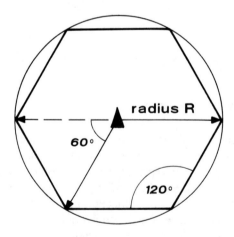

Figure 1.6 The regular hexagon fitted within a circle of radius *R*.

This brings us to the reason for multiple cells, frequency agility, and other features in the mobile station.

1. Other frequencies are needed in the adjacent cells.
2. How can the region of overlap between cells, which is very apparent in Figure 1.5, be managed?

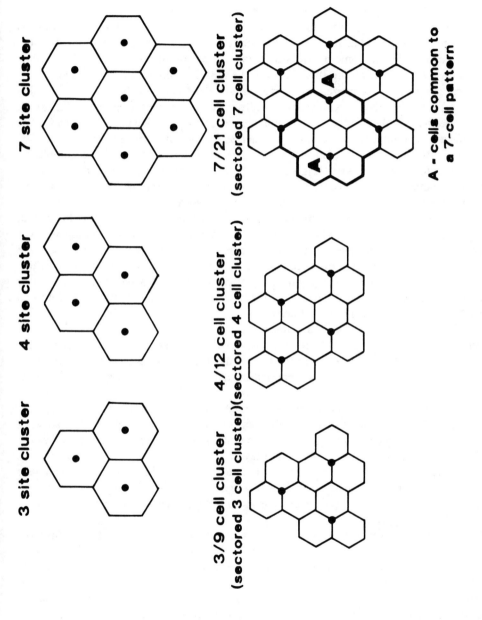

Figure 1.7 Cell repeat patterns. The base station locations (sites) are indicated with black dots. Several cluster sizes are shown.

3. When can the same frequencies be used again in these pattern arrangements, given that there is a limit to the number of assignable frequencies?

4. The cell sizes will need to be varied to meet subscriber demand.

The first point emphasizes that a cellular radio system will need a block allocation of frequencies. Whether the voice and signaling (control) are managed on a time division, a frequency division, or some hybrid arrangement does not matter; more bandwidth than the bandwidth required by the subscribers in the area of a single cell must be allocated. The first point also emphasizes that a mechanism must exist for handing over (called *handover*) the subscriber's radio connection from the frequencies used in one cell to the frequencies used in the next cell as the subscriber roams the service area. This is a feature essential to any cellular radio operation, whether very small cell sizes (*microcells*), average, or large cells (sometimes called *macrocells*) are considered. Much of the complexity within a cellular radio handset (Figure 1.2), the frequency synthesizer, and control and memory functions is there mainly to manage handover. The radiotelephone system must clearly have frequency agility to operate in a cellular system.

The second point is the question of the physical overlap of cell coverage in general, not a feature apparent in drawings of cellular radio plans. The mobile station will not know which cell it is in unless it undertakes a scan of the spectrum. Usually, only the base stations will know, and they must be able to communicate with each other, which is why there is a need for an overlay network of communication associated with the base stations indicated in Figure 1.5 and 1.7. This network, which can be regarded as a fixed network, can be realized by cable, microwave links, or even satellite links.

With the existence of such a connection, each base station can communicate with another, and they can decide between themselves when to take command of a subscriber moving through the region of overlap; this is the process of handover. The practical strategy used will depend on the system and other factors and is a fairly complex signal control problem.

The fact that several frequencies are used in just one cell in order to accommodate many subscribers, plus the fact that further sets of frequencies must be used in adjacent cells, raises the third point: the question of when the same frequencies can be used again. This is again a fundamental feature of cell layout strategy. See Figure 1.8, which shows seven groups of seven-cell clusters. In the middle of each is a base station that can have the same frequency set. The distance between these centers is called the *reuse distance*.

This reuse feature is another cellular radio principle and is critical to the overall design of actual systems.

Returning, however, to Figure 1.7, the reuse distance can be spotted in the more compact lower cluster of cells, which are again basically cluster sizes of 3, 4, and 7. However, the cells are now sectored, with the base stations positioned at a common corner of the cells and radiating in 120-deg sectors, as indicated in Figure 1.6. The cells now divide up into 3/9, 4/12, and 7/21 cluster sizes. The reuse cells have now disappeared outside the diagram, together with any likelihood of reusing the same cell frequency set, because the cluster sizes are now, of course, 9, 12, and 21, respectively [9].

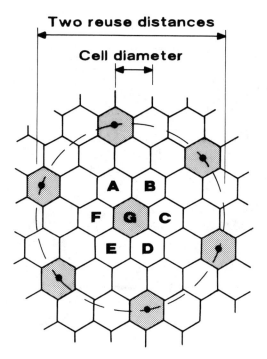

Two reuse distances

Cell diameter

A B
F G C
E D

Frequency sets A,B,C . . . ,G

Figure 1.8 A set of seven-cell clusters showing surrounding cells that can have the same operating radio frequency (RF) set because they are separated by the reuse distance.

Finally, we need to note that cell sizes need not be the same. Figure 1.9, for example, shows an initial cell layout plan for London, by the Vodafone Company, operating the Total Access Communication System (TACS). The cells are subdivided into smaller cells towards the middle of the city to permit management of a higher density of users. (The dotted lines represent motorways, such as the M25 ring road.) Also shown, shaded in, are some representative cluster sizes, namely 4, 7, and 12, for descriptive purposes only.

Cluster size N determines the frequency plan of the allocated radio spectrum; but as shown below, N is determined by the acceptable cochannel interference from the reuse cell frequencies.

Another feature of cell clusters is an electronic marking strategy. A mobile subscriber needs to recognize which cluster he or she is in, and a digital color code signal is sent to all mobiles by all the base stations in the same cluster. In a similar manner, supervisory audio tones ((SAT) in analog) or digital access grant channel messages (in digital), are found in order to establish the cell being used by the radiotelephone.

As mentioned, the cell cluster size affects the cellular radio plan. Thus, if one focuses on the popular seven-cell cluster arrangement of Figure 1.8, the allocation of

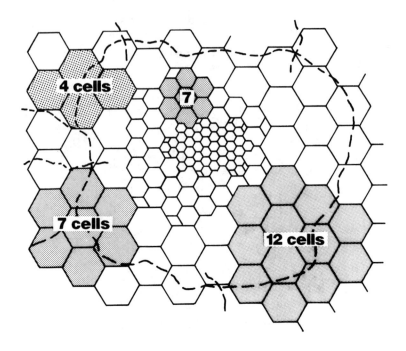

Figure 1.9 A cell layout plan typical for a city, drawing attention to cell sizes and possible cluster sizes.

frequencies into seven sets is clearly required. Therefore, if 210 frequencies (channels) were available, this would mean only 30 channels per cell could be assigned, of which one or more would need to be a control channel. Hence, 14 control channels, for example, would be needed for a seven-cell cluster. For duplex operation, this pattern would be repeated in the two allocated frequency bands (i.e., 28 voice channel pairs per cell).

The other aspect is the average distance to the surrounding cells using the same frequency set. This is the *mean reuse distance (D)*. Because of the geometry of hexagonal cells, D is related to the cell radius R, and the ratio of D to R, called the *reuse ratio*, is a function of cluster size, namely

$$\frac{D}{R} = \sqrt{3N} \tag{1.1}$$

For a seven-cell cluster with 2-mile radius cells, the repeat centers would be 9.2 miles away. For example, the distant, but similar, on-frequency transmitter will cause cochannel interference with a mobile in its rightful cell. For a seven-cell cluster, there would be up to six such interferers, as illustrated in Figure 1.8.

Assuming an inverse fourth-power distance propagation law (discussed below), an approximate value of the carrier-to-interference ratio (CIR), from Figure 1.8, is

$$\frac{C}{I} = \frac{C}{\Sigma 6I} = \frac{R^{-4}}{6D^{-4}}$$

Using the D/R ratio,

$$\therefore \frac{C}{I} = \frac{1}{6}(3N)^2 = 1.5N^2 \qquad (1.2)$$

This shows that the CIR is a function of the cluster size; it is designated C_i. For example, suppose $N = 7$, then $C_i \le 73$, or 18 dB. This appears adequate, but two other factors may need to be taken into account.

1. *Adjacent channel interference* from channels in neighboring cells. This is worse in small cell clusters and heavily used cells.
2. *Multipath fading* (i.e., irregular propagation), which may weaken C as compared to I.

It is useful to plot the number of channels per cluster and the C_i ratio versus cluster size N and note the step function nature of the result, shown in Figure 1.10.

The number of radio channels in any one cell is worked out by dividing the initial 300 channels, minus 21 control channels, as used in some cellular systems, by the number N. The C_i ratio is worked out using (1.2) and converting to decibels.

The interesting feature of Figure 1.10 is that only step function increases or decreases in cell layout are possible. The seven-cell pattern is most popular in analog cellular systems because it generally achieves an acceptable C_i ratio, but it means that only 39 channels (out of 300) are available for traffic service in each cell.

Sectorization can lead to smaller cell sizes because of directivity and better front-to-back ratio performance at cell sites, but it actually decreases the number of radio channels available because the number of individual cells in a cluster increases by three (see Figure 1.7). Overlaid, or umbrella, cells are sometimes used at centers of heavy subscriber operation, but jumping to a four-cell layout plan and 69 voice channels per cell, as suggested in Figure 1.10, requires a different C_i strategy. The new digital cellular operational networks achieve this by time division of the channel's operation, and as a result more efficient four-cell pattern layout is possible, as discussed in Chapters 6 and 7.

1.2.4 Cellular System

The basic concepts of frequency planning and reuse and the control of cochannel interference are applicable to all radio services. What is different with the cellular system is that the individual base stations are interconnected to form a system offering continuous

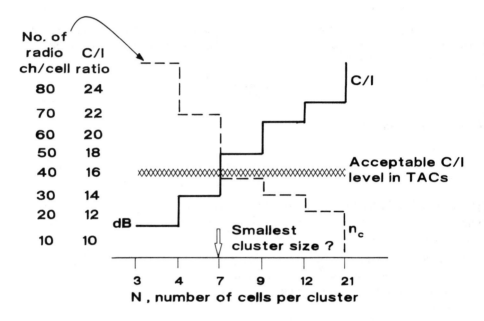

Figure 1.10 Number of radio channels per cell (out of 300) and the CIR (decibels) plotted against cell cluster size *N*.

coverage. There are two key features of the cellular system that make this possible: mobile location and in-call handover.

Mobile Location

When an outside call is received for a mobile station, the call has to be routed to the cell in which the mobile is located so the call can be connected. One way of finding the mobile would be to transmit a calling message (page) for the mobile on every cell site in the network. Clearly, with several hundreds of cells and many thousands of mobiles operating, the signaling capacity required would be too high. Instead, the cellular network is split up into a number of location areas, each with its own area identity number. This number is then transmitted regularly from all base stations in the area as part of the system's control information. A mobile station, when not engaged in a call, will lock onto the control channel of the nearest base station upon switching on and, as it moves about the network, will from time to time select a new base station to lock onto. The mobile will check the area identity number transmitted by the base station, and when it detects a change indicating that it has moved to a new location area, it will automatically inform the network of its new location by means of a signaling interchange with the base station. In this way, the network can keep a record (registration) of the current location area of each mobile, and, therefore, it will only need to call the mobile within that area.

In-Call Handover

When a mobile station is engaged in a call, it will frequently move out of the coverage area of the base station it is communicating with, and unless the call is passed on to another cell, it will be lost. There are several methods of achieving this process of handover, depending on the technology. In an analog cellular system, there is network-controlled handover; that is, the system continuously monitors the signals received from mobiles engaged in calls, checking on signal strength and quality. When the signal falls below a preset threshold, the system will check whether any other base station can receive the mobile at better strength. If this is the case, the system will allocate a channel for the call on that new base station, and the mobile will be commanded by a signaling message to switch to the new frequency. The whole process of measurement, channel allocation, and handover may take a few seconds to complete, but the user will only notice a brief break in conversation as the handover itself is carried out.

Effective and reliable handover is not only highly desirable from the subscriber's point of view, it is also essential in the control of cochannel interference and thereby the maintenance of the cell plan, particularly as the cell size is reduced. A mobile operating in a nonoptimum cell will, in effect, be operating outside the cell designated for that area; the cell boundary will have been altered beyond its planned limit, and this will give rise to levels of cochannel interference above that planned for the adjacent cells.

A further means of controlling cochannel interference is that of mobile power control. As long as the base station is receiving a signal of adequate strength from a mobile, there is no need for the mobile to be transmitting extra power, so the base station can command the mobile to reduce power by sending a signaling message. Clearly, by reducing a mobile's power, the likelihood of its causing interference is also reduced, helping to control interference levels. The radiated power of at least one channel of the base station is kept constant, however, since this defines the cell size.

In the newer digital cellular technologies (Chapters 6, 9, and 10), mobile-assisted handover is possible. The mobile is able to measure for itself the signals from adjacent base stations in nonactive time slots, report back to the network, and aid in the decision to execute handover.

1.3 THE FIXED SUPPORTING NETWORK

The fixed supporting network associated with a cellular radio scheme has several fundamental tasks [2–9].

- It connects all the base stations together for communication signals and messages to and from subscribers operating on the network.
- It provides switching centers so that traffic can be directed around the network. The centers are called mobile switching centers (MSC). An MSC does not have to be associated with every cluster of cells, but is usually sited in population centers. As

a cellular radio network grows (i.e., the number of subscribers increases), the MSCs have to begin handling a very large amount of traffic. What could have begun as a fairly small switch (100×100 cross point (line-to-line), for example) must expand to a full-capacity fixed telephone exchange.

- Full intermeshing of MSCs becomes very costly, and therefore a second tier of overlay transit switching centers (TSC) is now common in cellular networks, giving a hierarchical network also found in the fixed PSTN.
- A home location register (HLR) and a visitors location register (VLR) will be associated with each MSC, although the registers themselves need not be physically associated with the location of the MSC, since the fixed network gives full connectivity.

A conceptual diagram of the supporting fixed network is shown in Figure 1.11. In practice, many more MSCs and especially base stations will exist, but all the main components referred to previously are shown. In addition, a network management center, nowadays referred to as the *operations and maintenance center* (OMC), is indicated. In general, there will be only one such center for each network. Its role is, as the name implies, network management, which includes ensuring that these tasks are performed:

- Fault conditions are recognized.
- Fault diversion strategies can be implemented.
- Extra traffic routing is supervised.
- Interference conditions are recorded.
- Maintenance programs are run.
- Subscriber base and income are monitored.

The registers also provide authentication of a subscriber attempting to use the network. Authentication is usually associated with the HLR, but is shown as a separate center (AuC).

All equipment used in the network, whether portable or a vehicle-mounted phone, carries an electronic identity number—the mobile identity number (MIN). This is usually a ten-digit number programmed into the mobile on registering by an agency known as a service provider. Meanwhile, an equipment identity register (EIR) checks out the status of the subscriber identity number. These units are all shown in Figure 1.11. The time of use of the network is also recorded at a billing center. In general, the charge made is independent of the physical distance between the two users of the same network, since even if the callers were in adjacent cells or in the same cell, the full cellular fixed network is really put into use.

Finally, the fixed network will have a trunk connection to the existing fixed network. This provides the connectivity between fixed subscribers and mobile subscribers. The fixed subscriber will use a mobile identification number to call; the mobile subscriber will use the usual area code number. In both cases, billing is made at the mobile tariff rate.

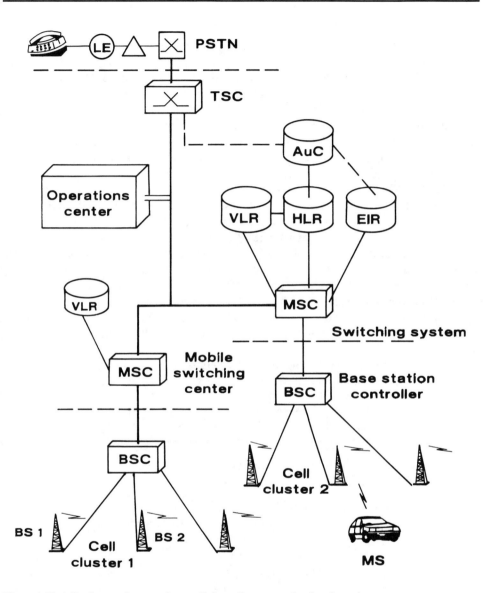

Figure 1.11 A fixed network supporting a cellular radio system, showing the major components.

1.3.1 Numbering Plans

The numbering plans associated with the fixed network (PSTN) and the mobile equipment identification plan are quite complicated.

The original numbering plan for the international telephone network is described in Consultative Committee for International Telegraphy and Telephony (CCITT) Rec. E.163. In this scheme, each country or zone is assigned a country code (CC) of one, two, or three digits, with a maximum overall international number length of twelve digits (excluding access prefixes such as 0, 00, and 010, etc.). This arrangement served until the early 1980s, when consideration was given to numbering plans for the Integrated Services Digital Network (ISDN). It was recognized that the numbering for the ISDN would evolve from PSTN numbering, so the numbering plan for ISDN is a development of the existing E.163 recommendation.

"The numbering plan for the ISDN era," is known as CCITT Rec. E.164. The main principles of E.163 remain, but an increased international number length of fifteen digits is allowed. This allows the network destination code (NDC) to cover the E.163 trunk code and also identify special networks. E.163 permitted up to two digits to determine the international route; this gave a variable maximum number analysis of three to five digits, depending on the length of the country code, whereas in E.164 this is fixed at a maximum of six digits, including country code. A three-digit country code is permitted within recommendation E.164, which is drawn as Figure 1.12.

A further recommendation, E.165, details the arrangements for the implementation of the E.164 plan and specifies the date of 31 December 1996 for bringing the E.164 into effect.

The identification plan for mobile subscribers is contained in CCITT Rec. E.212, and the allocation of mobile station roaming numbers is defined in E.213, shown overlaid

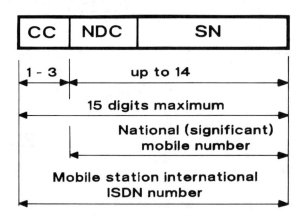

CC : Country code
NDC : Network destination code
SN : Subscriber number

Figure 1.12 The E.164 number structure and mobile numbering structure.

on Figure 1.12. The number consists of the country code of the country in which the mobile station is registered, followed by the national (significant) mobile number, which consists of a network destination code (NDC) and a subscriber number (SN).

The mobile identification code is actually limited to a ten-digit number (not the maximum of fifteen) made up of the country code, the NDC, and the SN. These are better known in cellular radio as MIN numbers.

MIN 1 is the country code, but there is now to be a mobile country code (MCC), which will be three digits. The first digit of this part refers to the world numbering zone, although these zones are not the same as those used by the fixed network.

The MIN 2 number consists of seven digits and usually corresponds to the mobile unit's telephone number. CCITT Rec. E.213 states the following.

- The numbering plan should allow standard telephone charging and accounting principles to apply.
- Each administration should be able to develop its own numbering plan.
- It should be possible to change the international roaming identity without changing the telephone number allocated to the mobile.
- Roaming without constraints should be possible.
- This numbering plan refers only to interconnection with the PSTN (i.e., it does not apply to mobiles that are not interconnected).

The MIN number is loaded in the mobile as part of the number assignment module (NAM). This number also contains data pertaining to its valid HLR of operation and an electronic serial number (ESN). This number enables the network to recognize that the mobile station equipment being used is valid and registered.

1.4 PROPAGATION ISSUES

Radio propagation is clearly an essential element of any cellular radio service. Chapter 11 describes the present computing strategies for determining actual cell plan coverage and other matters from propagation models. However, this book necessarily focuses on the application issues; the physics and vagaries of VHF and UHF propagation are essentially accepted verbatim. Many specific texts, of course, deal with this subject and much detail can be obtained from sources of this type [10–12].

The three basic factors of significance are:

1. The range in general;
2. Actual propagation;
3. Multipath effects.

Any cell within a set of cells in a network needs to address these three issues for itself and for the surrounding cells. Chapter 11 is devoted to software methods for cell planning; however, some physical knowledge of where the signals go, or do not go, is clearly useful, and this section attempts to do this.

1.4.1 Mean Signal Level

In free space, radio signal amplitudes (i.e., the mean envelope level of the carrier) obey an inverse distance law, and the higher the frequency the greater the attenuation. It is usual to express the radio signal level, observed at distance d from a transmitter, in terms of its received power, expressed in decibel milliwatts (i.e., dBm). The loss of transmitted power (P_T in Figure 1.3 expressed as dBW or dBm), by the time it reaches the receiver, is termed the propagation loss L (defined in dB). It is usually written in the form

$$L = 32 + 20 \log f_{\text{MHz}} + 20 \log d_{\text{km}} \qquad (1.3)$$

This shows that at 900 MHz, for example, a signal will suffer 20 dB more attenuation than one at 90 MHz; the same goes for distance when going from 9 to 90 km, for example, and L obeys an inverse square law.

This propagation law applies quite accurately to signals from space and clear line-of-sight microwave link paths. Unfortunately, our cellular radio subscriber is usually a terrestrial being, and reflection of the radio wave from the terrain (whether land or sea) severely changes the propagation law, and in the first instance it is usually written as

$$L = 40 \log d_m - 20 \log h_T h_R \qquad (1.4)$$

This equation indicates three important changes to the determination of propagation loss, namely:

1. Much more rapid attenuation with distance—the fourth power law. In literature, the factor α is often found to represent the coefficient. The equation here says $\alpha = 4$, being the 40 log d, and we note that d is now measured in meters.
2. The height above ground (a flat earth scenario can be assumed in cellular planning, in general) of both the base station h_T (well sited) and the mobile station h_R (near the ground), both measured in meters.
3. The frequency of operation now no longer plays a part.

Equation (1.4) enables the calculation of the range in general of a cell with an omnidirectional base station antenna. The maximum range (in theory), R_{max}, is the distance when the base station transmitter power (dBW), less the above propagation loss (dB), for the range d, just equals the total receiving site noise (dBW). This would correspond to a carrier-to-noise ratio of 0 dB, somewhat below a more realistic ratio of, say, +10 dB.

1.4.2 Actual Propagation

Observations show that, apart from the ultimate effect of the earth's curvature over a long path (large radio cell), several terrestrial effects must be taken into account, which include:

1. Surface irregularities;
2. Line-of-sight obstacles;
3. Buildings and trees;
4. Mountainous areas.

The propagation loss equation is therefore amended to read

$$L = 40 \log d_m - 20 \log h_T h_R + \beta \tag{1.5}$$

β is the additional losses lumped together. If the *additional loss factor* β is a constant, then this equation says the radio cell (range equal to the radius) is circular. Since β and h_R can vary according to which direction the transmitter is viewed from because of the terrain, radio cells in general are not circular.

Formula (1.5) is not complicated except that we do not know β. Also, it appears to be independent of frequency. This is certainly not the case. All observations show that the radio coverage decreases with increasing frequency; that is, VHF fares better than UHF. Moreover, because the term β has hidden all the possible geological and urbanization factors, the propagation law is a complex matter.

Empirical model formulas are based on fitting a formula to measured data. The actual measured signal data are compared to data predicted by the plane earth equation alone and the difference (excess clutter factor) found. The final model is therefore composed of the plane earth equation plus a clutter factor, which is a best-fit equation based on the factors considered most likely to increase propagation loss. It must be recognized, however, that such models only predict the mean signal level and hide the observed variability. Moreover, one model cannot be expected to fit all conditions and all locations, but the International Committee on Radio (CCIR) of the International Telecommunications Union (ITU) has suggested a model, based on a long series of observations, known as the *CCIR empirical formula for urban areas*. The equation is plotted as received field strength in Figure 1.13 for various frequencies of operation and assuming the conditions indicated. By drawing the −107-dBm (1 μV/m) threshold line on the signal level scale, some idea of the radio range R_{max} can be read off the distance scale d, shown in kilometers.

The model above does not consider penetration of radio waves into buildings. Penetration certainly takes place, the *building penetration loss* being a complex function of frequency and window sizes, and so on. A 10-dB reduction of signal level can generally be expected.

1.4.3 Multipath Propagation

The multipath situation in mobile radio is caused by reflection and scattering from buildings, trees, and other obstacles along the radio path and gives rise to much variability in the mean signal level. Radio waves arrive at a mobile receiver from different directions and with different time delays, and are made up of a possible direct ray, ground-reflected

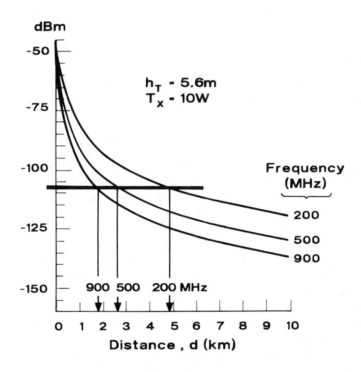

Figure 1.13 Signal strength versus distance according to CCIR recommendation for a base station height of 5.6m, $P_T = 10W$ and $h_R = 1.5m$. Different frequencies are shown.

rays, and other possible scattered rays. These combine vectorially at the receiver's antenna to give the resultant signal, which depends on path length differences that exist in this multipath field. Also, as a vehicle-borne or handheld receiver moves from one location to another, the phase relationship between the components of the various incoming waves change, so the resultant signal changes. Also, whenever relative motion exists, there is a Doppler shift of the frequency components of the received signal.

Characterizing the mobile radio channel is therefore a complex task (see [10–12]). Two cases can generally be identified. First, the case in which the signals occupy only a narrow bandwidth. *Narrowband* means that the spread of time delays in the multipath environment is sufficiently small so that all spectral components within the transmitted message are affected in a similar way. *Time delays* refers to the time of travel at the speed of light from the transmitter (base station, say) to the mobile, and the different rays will clearly vary according to the diversity of the path.

In the narrowband case, there are no frequency-selective effects, and the characteristics of the channel can be expressed in terms of their effect on any one component in the message; the carrier frequency is usually used. A more complicated form of characterization is needed to deal with wideband signals, the second case, when frequency-selective effects do occur.

The narrowband case occurs especially in urban areas because the mobile antenna is low, there is no line-of-sight path to the base station, and the mobile is often located in close proximity to buildings. Propagation is therefore mainly by means of scattering and multiple reflections from the surrounding obstacles, as shown in Figure 1.14. Because the wavelengths in UHF bands are less than 1m, the position of the antenna does not have to be changed very much to change the signal level by several tens of decibels.

This feature is known as *fading* and is observed as the position of the mobile station is changed. It can be witnessed in FM car radios and VHF PMR equipment. The signal appears to vanish at certain positions, but moving by a few meters brings it back again. The *signal envelope* takes on a standing wave pattern and is characterized by a log-normal probability density function. The mean signal level, about 3 dB below any peak value, is the one calculated in the previous section.

A receiver moving continuously through this field experiences a time-related variable signal, which is further complicated by the existence of Doppler shift. The signal fluctuations caused by the local multipath are now known as *fast fading* to distinguish them from the much longer term variation in mean level known as *slow fading*. A record of fast fading is shown in Figure 1.15. Fades of a depth less than 20 dB are frequent, but deeper fades, in excess of 30 dB, are less frequent. Rapid fading is usually observed over distances of about half a wavelength; therefore, at VHF and UHF a vehicle moving at 50 km/hr can pass through several fades in a second.

The multipath signal envelope is characterized by a distribution function of amplitude that follows the so-called Rayleigh distribution function, the reason narrowband multipath

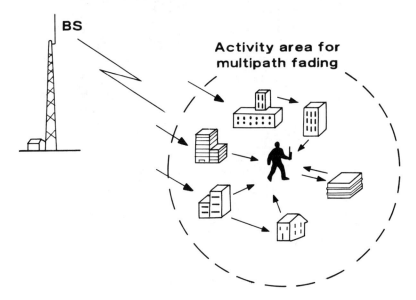

Figure 1.14 Illustration of radio propagation in an urban area.

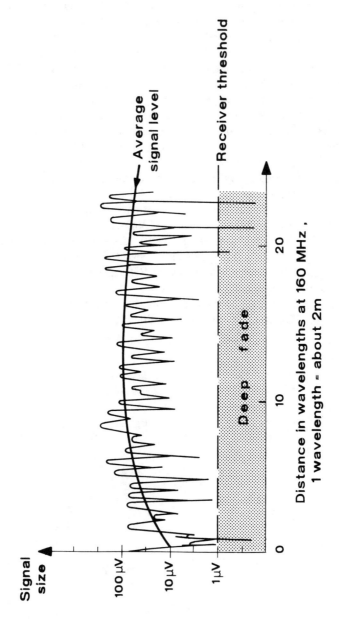

Figure 1.15 Illustration of a typical envelope pattern of a VHF signal received under multipath conditions.

is often known as *Rayleigh fading*. An important outcome of this result is that the observed received spectrum of a single carrier component from the transmitter takes on a specific power spectral density function distribution, represented in Figure 1.16.

Note that the limits of spectral deviation are $\pm f_d$, where f_d is the maximum Doppler shift frequency for a vehicle speed v.

$$f_d = \frac{v}{\lambda} \tag{1.6}$$

Component waves arriving from ahead of the vehicle experience a positive Doppler shift (maximum value $= f_d$), while those arriving from behind the vehicle have a negative shift.

It is worth noting that a speed of 60 km/hr, at 900 MHz, produces a maximum Doppler shift of 53 Hz; that is,

$$f_d = \frac{v \text{ m/sec}}{\lambda \text{ m}}$$

$$= \frac{60 \times 1{,}000}{0.33 \times 60 \times 60} = 53 \text{ Hz}$$

Also, a proportional change in frequency, or speed, will produce a proportional change in f_d.

This description applies to the envelope and phase variations of the signal received at a moving vehicle when an unmodulated carrier is radiated by the base station transmitter. On a more practical basis, the effects of multipath propagation of two or more frequency components within the message bandwidth need to be considered. If these frequencies

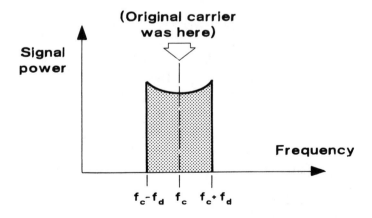

Figure 1.16 Power spectral density function of an RF carrier caused by Rayleigh fading.

are close together, then the different propagation paths within the multipath medium have approximately the same electrical length for both components, and their amplitude and phase variations will be very similar. This is the narrowband case. As the frequency separation increases, however, the behavior at one frequency tends to become uncorrelated with that at the other, because the differential phase shifts along the various paths are quite different at the two frequencies.

The extent of the uncorrelation depends on the spread of time delays, since the phase shifts arise from the excess path lengths. For large delay spreads, the phases of the incoming components can vary over several radians, even if the frequency separation is quite small. Signals that occupy a bandwidth greater than that over which spectral components are affected in a similar way will become distorted, since the amplitudes and phases of the various spectral components in the received version of the signal are not the same as they were in the transmitted version. The phenomenon is known as *frequency-selective fading* and the bandwidth over which the spectral components are affected in a similar way is known as the *coherence bandwidth.*

Figure 1.17 attempts to explain the situation. Here, we have three possible Rayleigh power spectral density functions, as in Figure 1.16, but now delayed in time relative to each other, since the principal scatters are markedly spaced in distance, those furthest away giving rise to the greatest excess path delay.

However, all the scatters come from the same source (say, the base station), and hence the signal power has to be distributed among the three cases. In practice, the

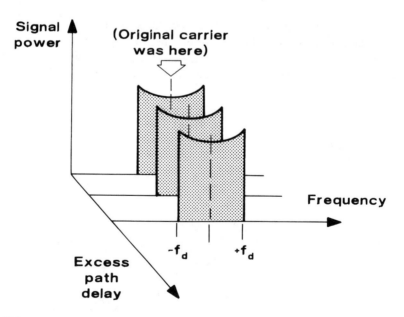

Figure 1.17 Power spectral density function for three principal path delays.

situation is much more like Figure 1.18. The Doppler shift of every component remains the same (i.e., f_d), but the spectral envelope will be a function of the particular environment. Also, it is only possible to describe a delay spread rather than a particular delay.

The coherence bandwidth is inversely proportional to the delay spread. Therefore, a large delay spread (in microseconds, say) signifies a small coherence bandwidth (on the order of, say, 100 kHz) and is manifested in the form of frequency-selective fading for wideband signals. This is a matter of concern to the Global System for Mobile Communications (GSM) technology (see Chapters 6 and 7).

1.5 MULTIPLE ACCESS STRATEGIES

This section assumes familiarity with modulation technologies in general, such as analog frequency modulation (FM), and digital shift keying modulations, such as frequency-shift keying (FSK) and phase-shift keying (PSK) [13, 14]. With analog FM radio, the signal bandwidth is allocated a maximum bandwidth within which the carrier deviates (e.g., 12.5, 25, or 30 kHz). With digital signaling, the most common method is fast frequency-shift keying (FFSK), which can convey data at 1,200 bps within a 25-kHz bandwidth after filtering. This is equivalent to 1 bit of data for every 2 Hz of bandwidth. More esoteric modulations aim at achieving nearer 2 bits of data for each 1 Hz of bandwidth

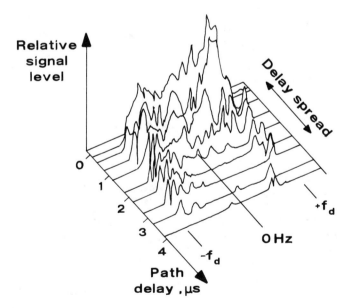

Figure 1.18 Typical scattered power profile in a suburban environment.

under multipath radio propagation conditions. This becomes important if the speech is digitized.

Figure 1.19 illustrates the problem. Analog radio telephony speech occupies, after careful filtering and compression, 3 kHz of bandwidth, which is then increased with FM to 12.5 or 25 kHz for radio transmission. The fixed telephone network has moved almost exclusively to 64-Kbps pulse-code modulation (PCM) (i.e., 8-bit encoded amplitude samples 8,000 times every second). This achieves toll-quality speech. Digital cellular is unable to accept such bandwidth extravagance, and linear predictive coding (LPC) techniques are used [15]. As will be explained later, the pan-European GSM digital cellular system operates with a protected bit rate of 22.8 Kbps. The method of modulation achieves 1.35 bits/Hz, so the system could operate in a bandwidth of 16.9 kHz. Figure 1.19 attempts

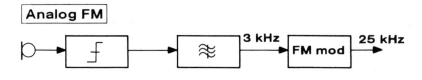

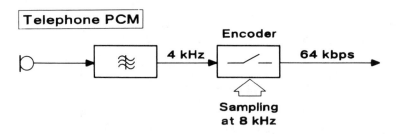

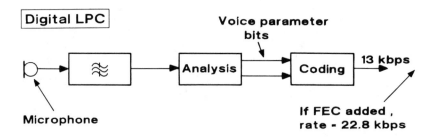

Figure 1.19 Analog FM, telephony PCM, and GSM speech strategy comparison.

to present these cases, but it must be made clear that the sketch is very much a simplification and represents one user at a time.

1.5.1 Frequency-Division Multiple Access

Cellular radio as a network does not need to specify the way individual subscribers are accessed by the network. Current analog systems use the familiar single-channel-per-user concept, known as *frequency-division multiple access* (FDMA). FDMA is the analog cellular standard. Figure 1.20 shows the frequency-division access arrangement in association with the use of a control channel.

The available spectrum is divided into channels A, B, C, D, and so on. The bandwidth of each channel is set by the transmitter emission mask (see Figure 1.21). It can be seen that the bulk of the transmitted (or received) power occupies less bandwidth than the allocated bandwidth. However, in the case of the analog TACS system, for example, although the allocated channel bandwidth is 25 kHz, the FM indexes used for voice and signaling exceed ±12.5 kHz and spill over into the adjacent channel. This conflict between allocated bandwidth and occupied bandwidth limits the channel allocation that can be used in the cell zones.

With FDMA, a single user will occupy each channel, and this is true whether the modulation is analog or digital. The signaling over the network is digital, being FFSK in most cases. Therefore, it would be quite straightforward to introduce digital speech, each channel becoming a continuous digital stream when in use. If the modulation method used has an efficiency of 1 bit/Hz, then, in principle, fully protected linear predictive encoded speech with a transmission rate of 22.8 Kbps could be carried by the 25-kHz channels (ignoring any signaling). There would then be a digital cellular radio scheme using FDMA.

The aim of going digital in second-generation networks is, however, apart from better compatibility with the fixed network supporting the cellular radio system, to have an alternative access method to achieve better spectral efficiency.

1.5.2 Time-Division Multiple Access

An alternative way of using the available spectrum is to let each user have access to the whole band for a short time (traffic burst), during which time the user transmits data as fast as he or she can [16]. The user's frequency allocation is shared with the other users who have time slots allocated at other times. Figure 1.22 shows a time-division multiple access (TDMA) arrangement where the spectrum is only partially allocated, each user group A, B, C, D, and so on, having a time slot allocation in a particular channel group, 1, 2, 3, and so forth.

This arrangement is sometimes known as *narrowband TDMA*. If all of the available spectrum, say the forward channel (a similar picture will apply to the reverse channel),

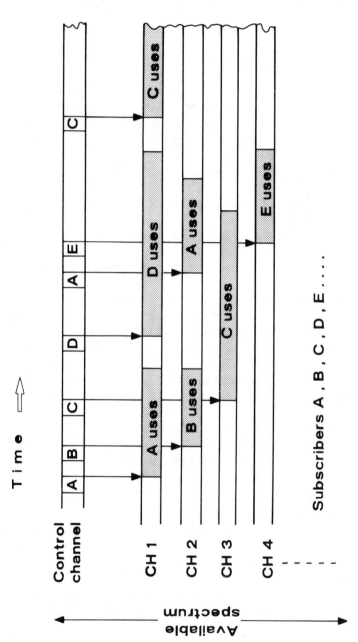

Figure 1.20 Demand allocation of voice channels on an FDMA basis.

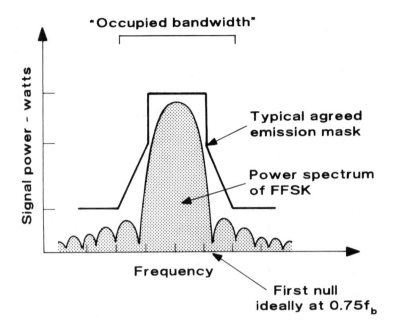

Figure 1.21 Transmitter emission characteristics.

was allocated to each user during his or her time slot, the situation would be wideband TDMA. Each user would have to transmit data at a high data rate in order to pass traffic. Alternatively, more leisurely frames could be set up within the same as those that apply to an existing FDMA scheme. The American and Japanese digital systems intend to begin with just three time slots per frame. The relationship between the number of subscribers per group frame length and data rates can be ascertained from Table 1.1.

Variation within the table data also arises due to the assumed speech coding rate and modulation efficiency factors.

The breakdown of frames into time slots with the individual message content is shown in Figure 1.23. Several variations of this arrangement are possible.

Here, N_F channels per frame are shown. The convention is to start with the number 0. Each time slot, belonging to an individual user, is made up of several parts, namely:

- Header contains bits for carrier recovery, bit timing recovery, unique words, channel identity
- C and S denotes control and signaling
- Traffic is the message part to the subscriber; for speech it would be encoded
- G denotes guard space to allow for time/distance propagation delay because of cell size

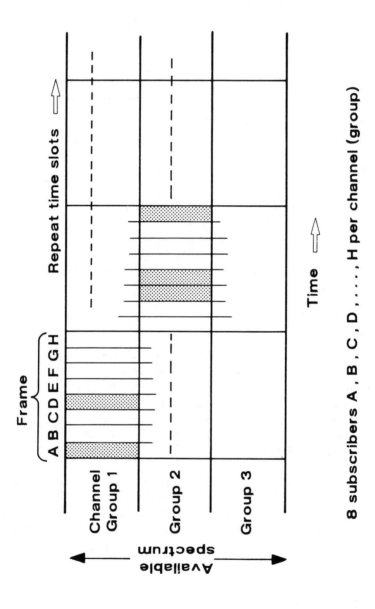

Figure 1.22 A TDMA mode of operation: each channel groups eight subscribers (A to H), who send (and receive) messages as bursts. The number of base station transmitters is equal to the number of groups.

Table 1.1
FDMA to TDMA Change and Parameters

System	TACS	ADC	GSM
Multiple access method	FDMA	TDMA	TDMA
Channels per carrier	1	3	8
Carrier spacing (kHz)	25	30	200
Number of channels	400	333	50
Frame period (ms)	No limit	40	4.6
Channel data rate (Kbps)	10	48	270
Number of user channels	400	999	400

Note: For the actual frequencies used for the forward and reverse channels, see relevant chapters. The table here assumes a 10-MHz spectrum available for each service in a frequency-division duplex mode.

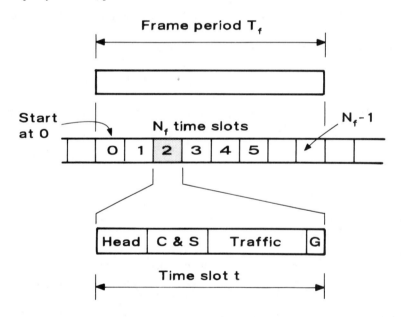

Figure 1.23 The frame, time slot, and message relationship in TDMA.

From this it can be seen that much of the activity of a TDMA system is taken up with what could also be called identification and authentication. The overall bit rate allocated to each user is only a percentage of that available to a single person per carrier scheme; on the other hand, no channels are uniquely allocated as control channels.

In TACS, or Advanced Mobile Phone System (AMPS), which uses FDMA, the control channels can be interpreted as reducing the spectrum efficiency by 21/300, or 7%. Also, there is the difficulty of not being able to allocate alternate channels. In TDMA schemes, the efficiency will depend on the number of nontraffic bits, as discussed in Section 1.6.3.

1.5.3 Other Technologies

The analog cellular system is known as a *first-generation system*. There are up to six incompatible analog cellular standards worldwide, which are described in detail in the following chapters. They have the underlying FM FDMA mode of operation. The aim of second-generation systems is (1) to go digital and (2) to have a more common global standard. The mode of operation allows TDMA, but matters like frame timing, method of modulation, and error correction procedures differ from one global system to another. The three main systems are GSM, Japanese Digital Cellular (JDC), and American Digital Cellular (ADC), which are described in later chapters.

At the time of writing (late 1992), these second-generation systems are in a highly advanced state of design, test, and trial status. Indeed, GSM is operational in some European countries and is being implemented in the Middle East and the Pacific region of Asia.

Third-generation systems are also being researched using methods such as code-division multiple access (CDMA) and controlled packet allocation. These considerations are briefly discussed in Chapter 15, but are, in general, well outside the scope of this text.

1.6 SPECTRAL EFFICIENCY

Spectral efficiency for a land mobile radio system [17] can be measured in two ways. The first is voice channels per megahertz per square kilometer. The number of radio channels per megahertz of the allocated spectral bandwidth is most easily appreciated for an FDMA system. The number of times the same channels can be used in a given land area depends on the cluster arrangements, as previously discussed.

This definition of spectral efficiency is based on fully specified parameters. However, spectral efficiency can also be defined in terms of the telephone traffic intensity, which can be supported by the network, giving the alternative definition: erlangs per megahertz per square kilometer. This measures the quantity of traffic on a voice channel or group of voice channels per unit time, which can be supported by the spectrum allocated over a region, which will depend on cell size.

1.6.1 Traffic Intensity and Grade of Service

The study of *traffic* is a well-established discipline in telephone engineering [18]. Telephone calls are made by individual customers according to their habit, needs, or in the conduct of business. The aggregate of customers' calls follows a varying pattern throughout the day, and telephony network equipment, sufficient in quantity to cope satisfactorily for the period of maximum demand, usually termed the *busy hour*, is a matter of public telephone operator planning. The basic factors involved in the nature of the demand are the *call attempt rate, call holding time, numbers of channels* (trunks or facilities), and *grade of service* (GOS).

The product of the first two factors is the *offered traffic*. It denotes the amount of time that a quantity of callers desires the use of facilities. A load that engages one channel (trunk) completely is known as an *erlang*: one erlang represents a circuit occupied for one hour; thus, 1 erlang = 1 call-hour/hour. The maximum traffic density in a single channel is one erlang, and the network attempts to achieve this status with the available equipment.

GOS is a measure of the probability that a percentage of the offered traffic will be blocked or delayed. GOS, therefore, involves not only the ability of a system to interconnect subscribers, but also the rapidity with which the interconnections are made. As such, GOS is commonly expressed as the fraction of calls failing to receive immediate service (blocked calls), or the fraction of calls forced to wait longer than a given time for service (delayed calls).

The assessment of the GOS provided to the user and the determination of facilities required to provide a desirable GOS are based on mathematical formulas derived from statistics and the laws of probability. They lead to telephone traffic formulas, which will not be described here (but see [18]).

An important result of telephone traffic formulas is that for subscribers making short calls over a common trunked network, the number of channels appears far higher than what actually exists (the so-called *trunking gain*). The same result applies to cellular, since there are several channels per base station. Typically for a GOS = 0.01 (1% of calls fail), the number of channels per base station appears to increase by a factor of 30 if a good number of channels is available.

1.6.2 Calculation of Spectral Efficiency

If the more specific definition is accepted (channels per megahertz per square kilometer), this implies that the efficiency can be defined as

$$\eta_s = \frac{\text{Total number of channels available}}{\text{Total } BW \text{ available} \times \text{cluster area}} \qquad (1.7)$$

Assuming a uniform cellular layout in clusters of N cells, each of which have a radius R, then it can be shown that an approximate result for η_s is the simple equation

$$\therefore \ \eta_s = \frac{1}{f_b} \times \frac{1}{C_i} \times \frac{1}{\pi R} \qquad (1.8)$$

Where f_b is the protected voice coding bit rate and C_i is the carrier-to-interference ratio.

The result is interesting in that it is the carrier-to-cochannel interference ratio that should be minimized; the number of cells in a cluster does not directly appear in the result.

Another factor that must be kept in mind, however, is the relationship between the forward error correction (FEC) and C_i ratio, since more correction may be needed to sustain a lower C_i ratio. The whole matter is complicated by the fading statistics of the wanted and interfering signals.

1.6.3 Access Efficiency

Provided the channels discussed above can be fully loaded with voice traffic, then the efficiency, in erlangs per megahertz per square kilometer, is almost equal to the factor η_s. To load up each channel, a multiaccess scheme must be employed, and it is assumed that there are many more users than channels. How efficient is this access facility? One approach is to multiply η_s by a term η_a, where

$$\eta_a = \frac{\text{Total time/bandwidth product, devoted to the voice channels}}{\text{Total time/bandwidth product, devoted to the system}} \tag{1.9}$$

$$\therefore \text{ Overall efficiency } \eta_o = \eta_s \times \eta_a \tag{1.10}$$

For *FDMA*, there is the need for guard bands between channels and for control channels. Clearly, $\eta_a \leq 100\%$

For *TDMA*, the efficiency will depend on how many bits in a time frame are dedicated to the message and how many are overheads. In narrowband TDMA systems, an FDMA breakup of the available spectrum also exists, as shown in Figure 1.22. For each channel group, however, there is

$$\eta_a = \frac{\text{Time slot duration for voice} \times N_F}{\text{Frame duration}} \tag{1.11}$$

where N_F is the number of time slots for voice transmission per frame.

If we take GSM (Chapter 6) as an example, and time slot period = 576.9 μs, frame period = 4.615 ms, number of users per frame = 8, we have

$$\eta_a = \frac{576.9 \times 8}{4,615} = 100\%$$

Unfortunately, much of each time slot is taken up with error protection bits, tail bits, training sequences, and a guard space, so a more accurate indication of the access efficiency comes from the alternative formula

$$\eta_a = \frac{\text{Voice channel bit rate} \times N_F}{\text{System (group) bit rate}} \qquad (1.12)$$

where voice channel bit rate = 13 Kbps and group bandwidth available = 270.833 Kbps

$$\therefore \eta_a = \frac{13 \times 8}{271} = 38\%$$

This apparent loss of efficiency is due to the fact that many of the actual time slot bits are given over to other activities, as just mentioned. On the other hand, the benefits of voice and data security are provided as part of the GSM service.

For narrowband TDMA, as was already explained, more spectrum is available in the other channel groups on the FDMA principle. Therefore, the access efficiency must also be modified by the term

$$\eta_a' = \frac{(\text{Bandwidth for a frame}) \times \text{no. of channel groups}}{(\text{Total } BW \text{ of system})} \qquad (1.13)$$

where bandwidth per frame = B_f, number of radio bands = N_c, and system bandwidth = BW.

$$\therefore \eta_a' = \frac{B_f \times N_c}{BW} \qquad (1.14)$$

Again, using GSM, $B_f = 200$ kHz, $N_c = 125$, and $BW = 25$ MHz (twice for duplex).

$$\eta_a' = \frac{200 \times 125}{25,000} = 100\%$$

so that no efficiency is lost through the narrowband TDMA format plan.

In TACS, $B_f = 25$ kHz, $N_c = 1,000$, and, again, $BW = 25$ MHz, so the access efficiency would again appear to be unity. As described, several channels are allocated to control channels between the different operators, so the access efficiency is actually less than unity.

1.6.4 Overall Efficiency

The overall efficiency [19–22] is the product of spectral efficiency and access efficiency, namely

$$\eta_o = \eta_s \times \eta_a$$

Because, as we found, η_a should effectively be 100%, we accept the overall efficiency as being mainly governed by η_s. The formula for η_s as given in (1.8) is an approximation, since it hides the loss of performance due to FEC. Also, in the case of narrowband TDMA, the access efficiency is not unity, again due to all the overheads in the speech coding algorithm, the time slot, and the frame itself, and represents quite a serious loss of efficiency. Nevertheless, the basic formula for η_s indicates the real limits of cellular radio. These are:

1. The voice communication bandwidth or bit rate. Here, digital technology is now drawing level with analog.
2. The tolerable carrier-to-cochannel interference. A 6-dB change here can represent a 40% improvement in efficiency. This is the aim of digital technology.
3. Any minimizing of the cell size will show real benefit. Power control of the base station and mobile transmitter by digital control makes for smaller cell geometry.

1.7 CELLULAR SUMMARY

Before we set out to explore the many cellular systems in operation and being installed worldwide with the guidance of specialists, it is perhaps useful to introduce a table that summarizes their features. Thus, Table 1.2 lists the parameters of the most widely available cellular systems. In addition, Table 1.3 gives their acronyms and the chapters in which they are discussed.

Not all these names appear in the table. Some are being overlaid on existing analog systems (e.g., ADC and JDC); others are interleaved, such as NAMPS (a 12.5-kHz version of AMPS). The second column in the table indicates the year in which the particular service started. The third and fourth columns give the particular frequencies allocated to the service, at least in the country where it began. There are two bands, normally separated by 45 MHz. The higher band is used for the so-called forward path, the lower band for the return path. The fifth column indicates the modulation method used for voice transmission and the method of multiaccessing. The particular scheme will determine the channel bandwidth allocated to both the base and mobile transmitter. Dividing this bandwidth into the allocated spectrum gives the number of channels in column 9. Note how the TDMA scheme (GSM) has to be multiplied by the number of users per frame. The data rate, or signaling, in column 7 is a more complicated issue, so it really only draws attention to a characteristic feature of the systems. The cluster size in column 8 is also a characteristic particular to the various systems, and points concerning this will be found in the relevant chapters. Handover strategy, as briefly described earlier, is normally controlled by base stations in analog systems. In the more recent digital systems, mobile-assisted strategies are employed. Finally, the last two columns give some idea of the market size, at least prior to 1992. The concept of truly global cellular is stranded on the rock of frequency allocation, but the size of cellular is, of course, quite staggering. Each system will have its own extensive fixed supporting network, which must not be overlooked when considering any one radiotelephone.

Table 1.2
Parameters of Current Cellular Radio Systems

System Name	Start Date	Frequency Band (MHz) BS$_{Tx}$	MS$_{Tx}$	Mode and Modulation	Channel Bandwidth (kHz)	Data Rate (Kbps)	Cluster Size	Channels	Handover System	Countries*	Subscribers* (millions)
NMT450	1981	463–467.5	453–457.5	FDMA FM ±4.7 kHz	25 20	1.2	60-deg sector	180 225	C/N at BS	14	15
AMPS	1983	869–894	824–849	FM ±12 kHz	30	10.0	7 12	832	C/N at BS	37	9
TACS (ETACS has extended BW)	1985	935–950 917–933	890–905 872–888	FM ±9.5 kHz	25	8.0	7 12	1000 1640	C/N at BS	21	2.5
NMT900	1986	935–960	890–915	FM ±4.7 kHz	25 12.5	1.2	7	1000 2000	C/N at BS	8	0.5
JTACS	1979	925–940	870–885	FM ±5 kHz	25	0.3	14	600	C/I value	1	2
JNTACS	1988			±2.5 kHz	12.5	2.4		1200 (2400)		9	
C-Net	1985	461–466	451–456	FDMA ±4 kHz	20	5.28		222	Propagation delay at BS	3	0.5
GSM (occupied by TACS)	1992	950–960 (935–950)	905–915 (890–905)	TDMA (GMSK)	200	271	4	600†	MS-assisted	24‡	N/A◇

* Approximate.
† 75 radio channels × 8 TDMA = 600.
‡ Based on a memoradum of understanding.
◇ N/A = Not available (because the system has only just started).

Table 1.3

Acronyms for Cellular Systems

Acronym	Definition	Chapter Discussed
AMPS	Advanced Mobile Phone System (U.S.)	Chapter 2
NMT	Nordic Mobile Telephone (followed by a number referring to the frequency band)	Chapter 3
TACS	Total Access Communication System (U.K.)	Chapter 4
ETACS	Extended TACS, offering more channels by additional frequency assignments	Chapter 4
JTACS	Japanese Total Access Communications Systems	Chapter 5
ADC	American Digital Cellular, a digital system overlaid on AMPS in the United States	Chapter 9
JDC	Japanese Digital Cellular	Chapter 10
C-Net	Cellular Network, specifically developed in West Germany. C is a later development	
GSM	Global System for Mobile Communications, specified by Committee of European Posts and Telecommunications (CEPT)	Chapters 6 and 7

REFERENCES

[1] Noll, M. A., *Introduction to Telephones and Telephone Systems*, Norwood, MA: Artech House, 1986.

[2] Young, W. R., "Advance Mobile Phone Service: Introduction, Background and Objective," *B.S.T.J.*, Vol. 58, 1979, pp. 1–41.

[3] Walker, J., ed., *Mobile Information Systems,* Norwood, MA: Artech House, 1990.

[4] Boucher, J. R., *Cellular Radio Handbook,* Quantum Publishing, 1990.

[5] Lee, W. C. Y., *Mobile Cellular Communications Systems,* New York: McGraw-Hill, 1989.

[6] Thrower, K. R., "Mobile Radio Possibilities," *J.I.E.R.E.*, Vol. 57, 1987, pp. 1–11.

[7] Beddoes, E. W., and R. I. Germer, "Traffic Growth in a Cellular Telephone Network," *J.I.E.R.E.*, Vol. 57, 1987, pp. 22–26.

[8] Hughes, C. J., and M. S. Appleby, "Definition of a Cellular Mobile Radio System," IEE Proceedings, Vol. 132, pt. F, Aug. 1985, pp. 416–424.

[9] Beddoes, E. W., "UK Cellular Radio Developments," *Elec. & Comms. Eng. J.*, Aug. 1991, pp. 149–158.

[10] Jakes, W. C., ed., *Microwave Mobile Communications*, New York: Wiley-Interscience, 1974.

[11] Lee, W. C. Y., *Mobile Communications Engineering,* New York: McGraw-Hill, 1982.

[12] Parsons, J. D., *The Mobile Radio Propagation Channel,* London: Pentech Press, 1992.

[13] Macario, R. C. V., ed., *Personal and Mobile Radio Systems,* London: Peter Peregrinus, 1992.

[14] Maral, G., and M. Bousquet, *Satellite Communication Systems,* New York: Wiley-Interscience, 1986.

[15] Xydeas, C., "Speech Coding," Chap. 4B in *Personal and Mobile Radio Systems,* R. C. V. Macario, ed., London: Peter Peregrinus, 1992.

[16] Calhoun, G., *Digital Cellular Radio,* Norwood, MA: Artech House, 1988.

[17] Hatfield, D. N., "Measures of Spectral Efficiency in Land Mobile Radio," *IEEE Trans. EMC,* Vol. 19, Aug. 1977, pp. 266–268.

[18] Boucher, J. R., *Voice Teletraffic Systems Engineering,* Norwood: Artech House, 1988.

[19] Murota, K., "Spectrum Efficiency of GMSK Land Mobile Radio," *IEEE Trans., Veh. Tech.,* Vol. 34, May 1985, pp. 69–75.

[20] Hummuda, H., J. P. McGeeham, and A. Bateman, "Spectral Efficiency of Cellular Land Mobile Radio Systems," *IEEE Veh. Tech. Conference,* Vol. 38, 1988, pp. 616–622.

[21] Lee, W. C. Y., "Spectrum Efficiency in Cellular," *IEEE Trans., Veh. Tech.,* Vol. 38, May 1989, pp. 69–75.

[22] Raith, K., and J. Uddenfeldt, "Capacity of Digital Cellular TDMA Systems," *IEEE Trans., Veh. Tech.,* Vol. 40, May 1991, pp. 323–331.

Part II
Analog Cellular Radio Principles

Chapter 2

Analog Cellular Radio in the United States

B. J. Menich

Motorola Cellular Infrastructure Group

AMPS is the current cellular air interface standard in the United States and the pillar of the analog cellular network. At the time of this writing, the system is dominant throughout North America and has achieved a moderate amount of penetration on other continents as well. With an installed worldwide subscriber base in excess of 10 million units, it is highly probable that AMPS will survive past the year 2000. A newer AMPS-like specification known as NAMPS has the potential to double, or triple, the number of analog subscribers in North America and provide some additional features now common to most digital cellular standards [1–3].

2.1 BACKGROUND

Before turning our discussion to AMPS, we will discuss briefly the history of some of the technology leading up to AMPS.

The immediate predecessor to AMPS is improved mobile telephone service (IMTS). IMTS owes its beginnings to all trunked radio systems that came before it and succeeds its predecessor (MTS) in that it offers automatic functions. IMTS is a full duplex system and is offered in two frequency bands: 150 MHz (known as *MJ*) and 450 MHz (known as *MK*) [4]. Spacing between the duplex frequencies is about 5 MHz and the channel spacing is 25 kHz. An IMTS cell site is usually located in a geographically high area with relatively high transmit power used to ensure a range of up to 25 miles. Because the mobile stations are allowed relatively high output powers (between 13W and 30W), one cell site location could serve an entire city. Some systems implemented remote receiving systems to boost the range and reception quality of the signal from mobile station to base station.[1] Thus, owing to the enormous coverage area and high transmit powers, IMTS is not a cellular service.

Signaling between the base station and mobile station takes place through the use of various inband audio tones. Signaling can take place either by turning a tone on and off, or by switching between two tones. The base station uses 2,000 Hz and 1,800 Hz to signal idle state and seizure state, respectively. The mobile station uses 1,336 Hz to signal disconnect, 1,633 Hz to signal connect, and 2,150 Hz as a guard signal.

Although IMTS is still used in the United States today, the small number of channels provided limits its availability to only a select few users in any city or town. In addition to the cumbersome use of spectrum, the signaling between the base station and mobile station was relatively ponderous because of the integration time of the tone detectors. A new, higher capacity signaling system was needed to support the paging rates that a highly available system would dictate.

Due to pent-up demand for land/mobile services accruing since the early 1960s, the U.S. Federal Communications Commission (FCC) invited industry proposals for a new land/mobile telephone system [1]. AT&T responded with a proposal whereby frequencies could be "reused" within a system of cells. Mobile stations would traverse the system of cells, *handing off* (changing frequencies) from cell to cell but keeping the connection into the PSTN intact.

2.2 AMPS

2.2.1 Requirements

In the creation of AMPS, Bell Telephone Laboratories cited such major system objectives as "efficient use of spectrum" and "widespread availability" as goals of the system [1]. Obviously, the major drawback of the IMTS systems was the inefficient use of allocated spectrum. Reusing frequencies provided the means for supporting a number of concurrent conversations far in excess of the number of voice channels derived from simply parceling the spectrum allocation. *Cell splitting,* or adding cells between existing cells and reusing frequencies, could theoretically continue for a long time before a system was completely interference-limited. In terms of "widespread availability," the cost of the technology involved in creating the subscriber equipment needed to be cheap enough to support sustained sales. It was hoped that future technological advances in miniaturization could be called upon to ensure penetration of lower price tiers in the subscriber market and thus perpetuate the service.

In the late 1970s, the FCC issued licenses in Washington, D.C., and Chicago, Illinois, for test systems to validate the AMPS interim specification. This validation phase preceded

---

[1]The receiver with the best signal wins out and is selected to provide the signal from mobile station to base station.

granting of construction permits for commercial systems by only a few years. By the mid 1980s, almost all of the top 20 markets in the United States were providing service.

2.2.2 Enabling Technologies

Several critical technologies came together simultaneously in the late 1970s to propel the cellular telephone industry forward. These technologies enabled small, relatively lightweight subscriber equipment to be manufactured cheaply. Some of the integrated circuit approaches allowed complementary metal oxide semiconductor (CMOS) technology to be applied to control circuitry, thereby reducing power consumption.

- The frequency synthesis function that was shrunk down to the scale of a single digital integrated circuit provided the means to do away with complex networks of crystal oscillators and tuning elements. In older style systems, these circuit networks needed to be replicated for every possible duplex frequency pair to be used.
- The microprocessor and microcomputer allowed complex control functions such as cellular call processing and handling of user keypad entry to be performed within a very small space.
- Application-specific integrated circuits (ASIC) provided a means of reducing the number of small- or medium-scale integrated circuits in order to implement many digital functions (e.g., FSK detection, microprocessor memory decoding, etc.).
- Surface-mount technology (SMT) applied to discrete circuit elements allowed for the shrinkage of linear circuitry blocks. This, of course, resulted in savings in weight and volume.

2.2.3 Air Interface

The AMPS air interface (RF) is currently specified by the American National Standards Institute, Electronic Industries Association (EIA), and Telecommunications Industry Association (TIA). The current version is known simply as EIA/TIA-553.

Figure 2.1 shows the frequency range that AMPS occupies. The base station transmit and receive bands are separated by 45 MHz. The channel spacing is 30 kHz, and each operator within a geographical area is allocated exactly half of the available channels (416) for control and voice. The B frequencies are reserved for the wireline operator and A frequencies for the nonwireline operator. The channels in the A', A", and B' bands are sometimes referred to as extended AMPS (EAMPS) channels, since they were added late in the specification stage and after AMPS commercial introductions. The modulation technique for the user voice is frequency modulation.

There are two classifications of AMPS channels. Control channels are used by the base station and mobile station to exchange information relating to call setup. Twenty-one channels are reserved in both bands for the implementation of control channels. Mobile stations always monitor the control channels when not in conversation state and

Mobile
Transmit

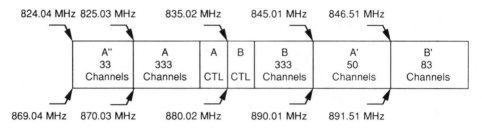

824.04 MHz 825.03 MHz   835.02 MHz   845.01 MHz   846.51 MHz

A" 33 Channels	A 333 Channels	A CTL	B CTL	B 333 Channels	A' 50 Channels	B' 83 Channels

869.04 MHz 870.03 MHz   880.02 MHz   890.01 MHz   891.51 MHz

Base
Transmit

Figure 2.1 AMPS band spectrum assignments. Note that 21 channels are always allocated for both the A and B systems.

lock to the strongest of these to acquire paging and overhead information. The traffic channels (TCH) are used by the base station and mobile station to bear information during the conversation state of a call.

Link continuity between the base station and mobile station in a TCH is accomplished via the SAT signal, which is added to the voice signal prior to the modulation of the RF carrier. There are three SATs: 5,970, 6,000, and 6,030 Hz. The signals are generated by the base station. The mobile station listens for a particular SAT and, after detecting it, regenerates it back to the base station. Both sides of the link use the SAT to know the transmit status of the other party during a phone call. Mobile station transmitters are disabled upon five contiguous seconds of SAT absence. Thus, a safety feature is built into the specification should there be an error in call teardown procedures. In addition, loss of the SAT at the receiving side may be used to mute audio to the user.

Signaling between the base station and mobile station takes place through either the signaling tone (ST) or the 10-Kbps FSK data signaling. ST is used exclusively by the mobile station to signal:

- Disconnect (1.8 sec of uninterrupted tone);
- Request to send dialed digits (400 ms of uninterrupted tone);
- Acknowledge a handoff order (50 ms of uninterrupted tone);[2]
- Alert (continuous tone, removal of tone in off-hook condition).

The 10-kHz FSK data signaling is used as the exclusive data transmission method on the control channels and for all order messaging on the TCHs. An NRZ bit stream is

[2]The terms *handoff* and *handover* are synonymous. The former is used consistently in the United States, whereas the latter is standard in Europe.

encoded into a Manchester code that allows the receiving station to track the phase.[3] All data are further encoded into a BCH (Bose-Chaudhuri-Hocquenghem) code of (48,36) on the control channel and (40,28) on the TCHs.[4] Figure 2.2 depicts a forward TCH message. Note that each data word is repeated 11 times over the length of the transmission (about 100 ms). This redundancy in the transmission increases the probability of successful reception by the mobile station receiver.

The control channel in AMPS uses the 10-kHz FSK signaling to communicate with mobile stations in a point-to-point or point-to-multipoint mode. The forward control channel carries a signal known as *busy-idle status* (BIS) to all mobile stations. This bit informs all mobile stations as to the status of the reverse control channel. Since the reverse control channel is multiple access (multipoint-to-point), BIS signals whether or not a mobile station is currently using the reverse control channel to communicate with the system. This mechanism reduces the probability of collision on the reverse control channel due to several mobile stations trying to access the system at the same time. Other information carried by the forward control channel includes:

- CMAC—the maximum power level that a mobile station may use when accessing the system on the reverse control channel. CMAC is set on a per-cell basis and may be different from cell to cell.

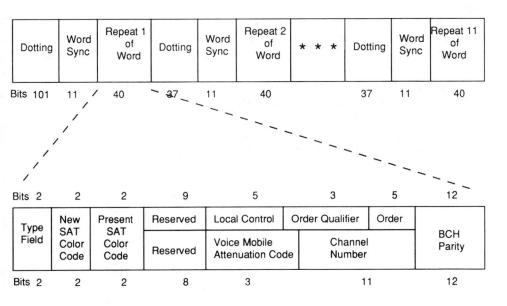

Figure 2.2 Forward voice channel signaling format.

[3]The Manchester encoding also prevents any dc bias from creeping into the signal via a long series of ones or zeros in the baseband data.
[4]These codes are well-known cyclic codes that allow multiple error detection and correction in channel coding applications.

- DCC—the digital color code of the base station. DCC is used by the mobile station in access attempts and page responses so that the base station may know that the current action by the mobile station is not the result of a cochannel control message.
- The maximum number of allowed attempts by the mobile station to seize the reverse control channel during an access attempt.
- Parameters relating to registration, such as the system identification (SID), which allows the mobile station to determine if it is operating in its home system or in a foreign system.
- Whether or not this particular control channel combines paging messages with access grant messages.
- Directed retry—used to force mobiles to make access attempts or page responses in other cells when the current cell cannot support a new call.

Mobile Station Specifics

Mobile stations in AMPS are categorized according to their maximum allowed effective radiated power (ERP). Each mobile can be commanded to a power level within an operating range of levels; however, the mobile may not radiate more power than is allowed for its class. The power control levels are separated by 4 dB each. Table 2.1 lists the range of power levels. Note that the prudent systems engineers always design their system (e.g., maximum cell sizes) for the lowest power class mobile that they expect will be used within the system.

Each mobile station is identified by two unique numbers. The first number is the MIN, which is a 34-bit binary encoded representation of the mobile station's 10-digit directory telephone number. This number is usually programmed into the mobile station when a subscriber begins service with a particular cellular operator. The second number is the ESN, which is a 32-bit binary number that is programmed into the mobile station

Table 2.1
AMPS Mobile Station Power Classes (watts)

Mobile Station Power Level	Nominal ERP		
	Class I	Class II	Class III
0	4.000	1.600	0.600
1	1.600	1.600	0.600
2	0.600	0.600	0.600
3	0.250	0.250	0.250
4	0.100	0.100	0.100
5	0.040	0.040	0.040
6	0.016	0.016	0.016
7	0.006	0.006	0.006

in the factory at the time of manufacturing and which uniquely identifies the equipment. The number is programmed in such a way that it may not be altered or erased. The ESN contains an 18-bit serial number and an 8-bit manufacturer code.

The *station class mark* (SCM) of a mobile station indicates the maximum ERP that the mobile can deliver, as well as the capabilities of the mobile regarding discontinuous transmission and operating frequency range (original 666-channel allocation/20 MHz, or extended 832-channel allocation/25 MHz).

Discontinuous transmission (DTX) is the term used to describe an operating mode in which the mobile station gates its transmitter output power as a function of speech activity on the part of the user. DTX may not be used by a mobile station unless it is allowed in the system. Two bits in the overhead message of the AMPS control channel indicate the DTX permissions in the system. When a mobile station is in the *DTX-high* state and is involved in a call with a cell site, it radiates at either the maximum amount of power it is capable of, the maximum amount of power for which it is allowed in that particular cell site, or the most recently commanded power. When it is in the *DTX-low* state, the mobile station radiates at a level 8 dB below the DTX-high state or at any level up to the DTX-high state. The choice of DTX-low transmission power mode is up to the service provider. The reduction in radiated power (about 16% of the DTX-high state in the 8-dB reduction case) allows a battery savings in proportion to the voice activity factor.[5] Earlier versions of AMPS mobiles completely disabled their transmitters during the DTX-low state. This was found to be somewhat of a nuisance, since it then became possible for a cochannel or co-SAT interferer to capture the receiver at the base station. When in the DTX-low state, the mobile station suppresses regeneration of the SAT. This is done as a signal to the base station receiver that knows no user voice is being transmitted and that the receive audio may be muted. Thus, the other user is spared from listening to static. The base station equipment may command the mobile station to momentarily go to the DTX-high state by using the audit order. This technique may be used in the determination of the correct reverse TCH signal strength and in the verification of the mobile SAT regeneration status.

2.2.4 Network

As depicted in Figure 2.3, the AMPS network makes use of the PSTN with the addition of MSC, base station equipment, and interconnections between these modules.[6] The intent is not to show the configuration for any one specific AMPS manufacturer. Instead, possible elements and possible interconnects are shown. Many infrastructure manufacturers offer

---

[5]The voice activity factor is the fraction of time during which a single party in a duplex conversation is actively engaged in speaking.
[6]Individual AMPS network configurations are always manufacturer-specific. This is due to the fact that AMPS specifies the air interface only. Note that AMPS was developed well in advance of systems like GSM that specified network elements and their interfaces.

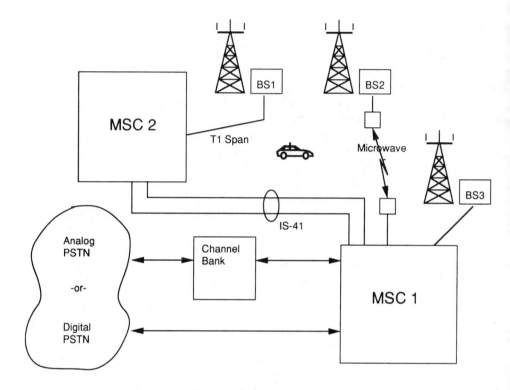

Figure 2.3 Typical AMPS network.

options that allow their customers to scale system configurations to meet the demand of their expected subscriber base. This scaling may manifest itself in the size and/or number of MSCs, base stations, and terrestrial circuits to the PSTN.

Multiple systems may be connected together via a standard known as IS-41. This standard specifies peer-to-peer communications and trunking between different cellular systems at the MSC level. Major attributes of IS-41 include intersystem handoff and call delivery (roaming). Call delivery allows a subscriber to be reached in a foreign system, providing that the subscriber has registered in the foreign system. IS-41 is a common standard which all MSC manufacturers adhere to and provides a common communications scheme at the network level that supports these cellular functions.

Several companies entered into the competition to provide infrastructure equipment to U.S. and Canadian operators. AT&T, Ericsson, Northern Telecom, and Motorola currently dominate in terms of subscribers served and voice channels provided.[7] The competition for infrastructure dollars is limited to large corporations like these because of the

[7]It is interesting to note that as of 1993 no Japanese or Korean companies have made significant infrastructure sales in North America. This is perplexing, since they compete fiercely in the subscriber unit market.

significant level of research and investment required. The infrastructure market is a demanding enterprise, necessitating tens of thousands of electronic components and more than a million lines of software.

2.2.4.1 Mobile Switching Center

The primary purpose of the MSC, shown in more detail in Figure 2.4, is to provide a voice path connection between a mobile station and a land line or between two mobile stations. The MSC is composed of a number of computer elements controlling switching functions, call control, data interfaces, and user databases. The MSC is usually based in a mobile telephone switching office (MTSO), which is either a wireline central office plant, or the environmental equivalent. Companies such as Ericsson and AT&T chose to base their cellular switching functions on existing product lines. Motorola chose to develop a new platform internally as well as buy switches from Digital Switch Corporation.

With respect to the PSTN, the MSC must be able to communicate control information, for example, common channel signaling and switch audio,[8] or PCM connections. Call control for PSTN connections usually takes the form of a software state machine implemented by microprocessors.

Cellular call processing tasks performed by the MSC vary between manufacturers. For land-to-mobile calls, the MSC must translate the PSTN-dialed digits into a MIN and initiate paging in the paging areas where the mobile station is suspected to reside. For

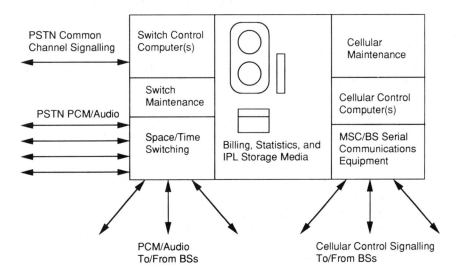

Figure 2.4 Typical AMPS mobile switching center.

---

[8]Even within the United States, not all central office telephone plants are digital-capable as of 1992.

mobile-to-land calls, the MSC must interpret the digits dialed by the mobile station and initiate a trunk seizure into the PSTN. Some manufacturers implement radio channel allocation and base station resource management at the MSC level, while others place execution of these tasks in the base station. On handoff, the MSC may be responsible for determining adjacent cells for measurement requests and implementing an algorithm to determine the best cell for the target of the handoff. The MSC then initiates and manages the sequence of events that terminates with the mobile station successfully arriving on the target cell.

Typical MSC call features usually available to AMPS customers include:

- Three-party conferencing,
- No-answer transfer,
- Speed dialing,
- Busy transfer, and
- Call waiting.

In addition to these features, the subscriber may also turn on or off any feature, typically by invoking a change service and manipulating the service with DTMF tones generated by the mobile station keypad.

The MSC may also implement the mobility management function. This function regulates the registration of mobile stations as a function of position or time. Registration may also be solicited by the system. The purpose of registration is to let the MSC know in which area of the mobile service area (MSA) the mobile station resides. This is important, since it allows the MSC to cut down on the number of cells used for paging a particular mobile. Tradeoffs are usually made between the size of paging areas and the amount of registration activity (impacts reverse control channel capacity as well as base station-to-MSC network capacity) on the part of the mobile stations.

Some manufacturers have implemented maintenance features for the entire cellular system at the MSC location. Idealistically, the cellular system strives to provide land-line quality service. To that end, the system must be fault tolerant and provide the following features in that area [5].

- Redundancy: Each component of the system must have a backup. If one fails, the alternate takes over.
- Detection: When a component fails, the fault must be made known so that it can be traced, corrected, and replaced.
- Isolation: To prevent a failed component from compromising the system operation, it must be automatically isolated from the rest of the system.
- Reconfiguration: In spite of component failure, the system must be capable of rapidly reconstructing itself to maintain operational continuity.
- Repair: Once repaired, the previously failed component must be restored to the system without interruption of system processing.

To accomplish these operations, the system is usually equipped with an array of sensors and diagnostic devices that may be controlled by the MSC or the base station.

Display of system alarms and diagnostics is usually provided at the MTSO. All major subsystems (audio, RF, switching, and computer equipment) are provided with alarms and diagnostic procedures.

Several different storage media may be available at the MSC location. These may include magnetic tape drives and disk drives. These media are used to store billing information and provide a means for loading the MSC initial program load (IPL). In addition, statistics relating to the health and performance of the cellular system may be collected. Examples of such statistics are shown in Table 2.2.

2.2.4.2 Base Station Equipment

The term *base station equipment* is usually reserved for radios and associated control electronics. However, the base station equipment may be expanded to describe the antennas, antenna mast, coaxial cabling, duplexers, equipment environmental housing, battery backup or emergency electrical generator, test gear, and other required equipment. Cell sites may be located on mountain tops, in buildings, in the middle of corn fields, or just about anywhere the service provider can procure property.

In general, cell sites are positioned according to a predefined frequency reuse pattern that attempts to constrain the cellular-induced self-interference (signal-to-interference ratio) to a known quantity (usually about 17 or 18 dB) or lower. Variations on this theme are a function of the type of antenna used to implement cells. For omnidirectional antennas, a twelve-site reuse pattern may be achieved. Seven-site and four-site reuse patterns may be achieved with 120- and 60-deg antennas, respectively.[9] A narrower beamwidth antenna "sees" less interference because of a more limited perspective, thus boosting the signal-to-interference ratio. Because of this, frequencies may be used in a tighter pattern, resulting in increased spectrum efficiency.

The chief elements of the base station equipment (see Figure 2.5) include control channel equipment, voice channel control equipment, locating receiver equipment, and

Table 2.2
Example System Statistics Collected at MSC

Call Statistics	Handoff Statistics	Maintenance Statistics
Attempted	Attempted	Cell site and MTSO Alarms
Completed	Completed	Number of out-of-service terrestrial circuits
Dropped	Handoffs to a particular target cell	Number of out-of-service radio circuits
Channel usage	Power levels	
Registration attempts		

[9]These reuse schemes are usually derived under assumptions of regular cell geometries and isotropic path loss from all cells. In reality, most system operators are not able to obtain the exact preferred site locations (thus violating the assumption of regular geometry), neither do their systems usually exhibit isotropic path loss.

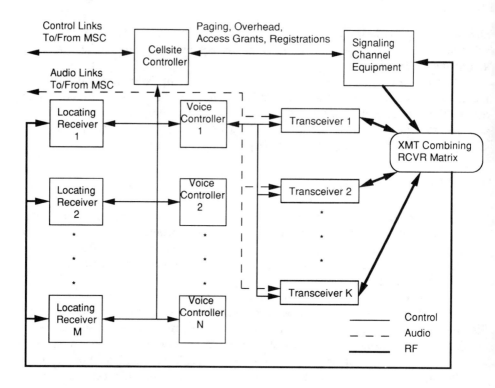

Figure 2.5 AMPS base station equipment.

computing machinery (cell site controller) to implement cellular control and fault manage-ment. A large part of the cell site equipment mass is dedicated to transmitter/receiver pairs for each duplex radio channel.

The cell site controller is responsible for coordinating all control functions at the cell site. Control functions may include management of cell resources (radio channels, locating receivers, etc.), some call processing functions, and fault management/initializa-tion. Typically, maintenance personnel may access the cell site controller through a terminal interface and exercise various testing routines for the purposes of calibration and testing of cell RF equipment.

The control channel equipment consists of one or more radio transmitter and receiver pairs, FSK modem equipment, and control electronics. Each cell site has a control channel that performs the functions of broadcasting general information about the system to all mobiles within the cell (point-to-multipoint), paging mobiles for mobile-terminated calls (point-to-multipoint), and sending voice channel orders (point-to-point). The control chan-nel sends page responses, registrations, and origination messages to the cell site controller. The cell site controller sends pages and access grants to the control channel equipment.

The voice controller equipment replicates a signaling tone detector and a SAT detector for each voice channel unit. An FSK modem may be replicated per unit or may be shared between several units (to reduce cost). The voice controller equipment may be responsible for processing some call processing information and detecting the need for handoff and power control. In addition, the voice controller may schedule scanning tasks for the locating receiver.

The locating receiver equipment may be used to scan local mobiles and handoff measurements requested by the MSC. The locating receiver may make signal strength measurements on any or all of the antennas at the cell site.[10] Typically, the locating receiver also verifies a particular mobile by determining which of the three possible SATs is being used. The accuracy with which the locating receiver estimates the signal strength of a mobile is of paramount importance to the correct functioning of the system. For accurate handoffs between cells, it is desirable to minimize differences in locating receiver measurements among all units in all cells. Note that some manufacturers implement local mobile scanning (i.e., estimating signal strength for mobiles being served by the cell) with the voice channel equipment, while others allow the locating receiver to perform this task.

Local cell site maintenance functions are usually carried out through a terminal interface with the cell site controller or through switches and knobs on the front panels of circuit boards and modules. An example of a common base station maintenance procedure is the retuning of channels because of cell splitting[11] or reallocation of frequency groups. At periodic intervals, cell site technicians may calibrate signal strength circuits in locating receivers, test transmit power levels of control and voice channels, and perform SINAD tests on receiver equipment.

The base station is usually equipped with a site-specific database that controls mobile station actions on the control channel and the voice channel. In addition, the database may also contain information regarding the site itself (e.g., number of voice channels equipped, number of locating receivers equipped, omnidirectional or sector configuration, etc.). Database information may either be downloaded to the cell site from the MSC or entered into it from the maintenance terminal. Typical Cell Site Parameters include:

- System ID Number,
- DTX indication,
- CMAC,
- Cell equipage,
- Registration,
- Control channel overhead information,
- Handoff thresholds,
- Power control thresholds,
- Voice channel, control channel, and scan receiver calibration data.

---

[10]One simple technique is to measure the peak-to-peak voltage of the carrier envelope and logarithmically compress this measurement to decibel microvolts.

[11]*Cell splitting* refers to the act of increasing system capacity or coverage by adding new cell sites to the system between already existing cells.

2.2.4.3 Network Connections

Besides the equipment that makes up the MSC, base station equipment, and mobile station, various other pieces of equipment are required to complete the network. Channel banks, T1 spans, and microwave links are used to provide communications pathways between the MSC and PSTN, and between the MSC and base stations. Some service providers use microwave networks in various configurations (star, daisy chain, etc.), bouncing signals between sites. Other service providers are restricted from using microwave links due to local environment zoning restrictions or a lack of spectrum. Thus, T1 circuits are used to provide links between MSC and base station, as well as PSTN connections. A technique to cover a large area with cell sites and then connect them back to the MTSO is to use satellite transponders in geosynchronous orbit.

2.2.5 Selected AMPS Call Processing Procedures

The AMPS specification does not give details on call processing procedures from the network perspective. Rather, details are given that specify, for example, what the mobile station or base station will do, how much time it has to do it, when it receives a particular message or when a timer expires. In this way, network call processing procedures may be derived from following reactions to message inputs and expiration of timers. The following are examples of call procedures that might take place in an AMPS system.

2.2.5.1 Mobile-Originated Call

Prior to the placement of a call, the mobile station monitors the 21 possible control channel frequencies in its assigned band. The purpose of this monitoring is to find the control channel with the strongest signal and lock onto that signal to watch for overhead information and paging. When the user originates a call, the mobile station will send in an FSK digital message on the reverse control channel (provided that the BIS indicates that the reverse control channel is currently idle) with the dialed digits entered, the MIN, and ESN. This activity is shown in Figure 2.6.

The base station reverse control channel receives the origination message and forwards the message (via the cell controller) to the MSC. The MSC receives the origination message, verifies the MIN/ESN combination, initiates seizure procedures into the PSTN, and allocates a TCH at the cell site that will bear the call. The MSC then sends a TCH assignment message to the base station with the identity of the TCH.

Upon receiving the TCH assignment message, the base station TCH keys its transmitter and begins sending the SAT. At the same time, the forward control channel sends an assignment order to the mobile station containing the frequency and the SAT of the TCH that will bear the call and the MIN that identifies the mobile station.

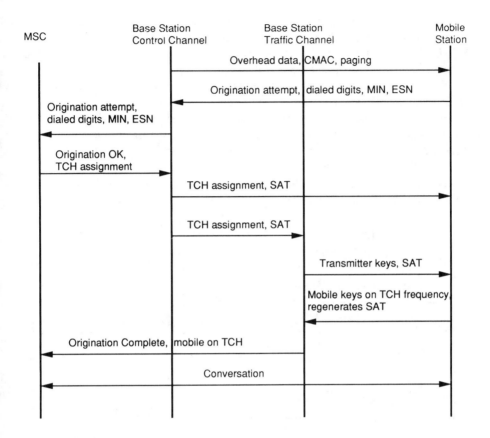

Figure 2.6 Mobile origination.

The mobile station receives the assignment order and retunes its frequency synthesizer and monitors for the designated SAT. Upon SAT confirmation, the mobile station keys its transmitter and regenerates the SAT to the base station.

The base station detects the regeneration of the SAT by the mobile station and sends an "origination complete" message to the MSC. The MSC then connects the PSTN trunk to the TCH trunk. Conversation may now take place, since both the base station and mobile station have reached the same state.

2.2.5.2 Mobile-Terminated Call

As in the mobile-originated call case, the mobile station monitors the 21 possible control channel frequencies in its assigned band and locks onto the strongest of these.

A user on the PSTN side of the system initiates a call by dialing a seven-digit number assigned to a mobile station. The PSTN routes the call to the cellular MSC. The MSC checks the paging area where the mobile station last registered and broadcasts a page to all base stations in that area, as depicted in Figure 2.7.

The base station receives the page message and transmits it on the forward control channel. The mobile station receives the page message while monitoring the control channel. The mobile station compares the MIN in the page message with the MIN stored in its read-only memory (ROM) and transmits a page response (provided that the BIS indicates that the reverse control channel is currently idle) containing the ESN and MIN on the reverse control channel.

The base station forwards the page response to the MSC. The MSC receives the page response message, verifies the MIN/ESN combination, and allocates a TCH at the

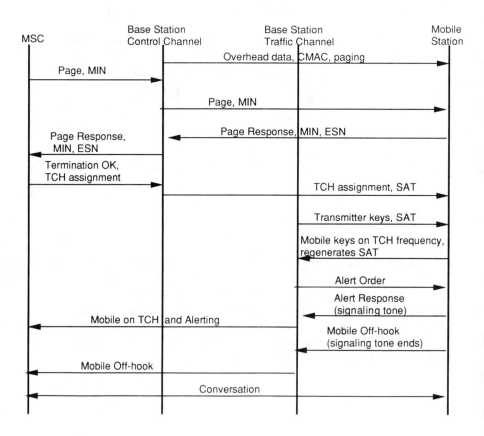

Figure 2.7 Mobile termination

cell site that will bear the call. The MSC then sends a TCH assignment message to the base station with the identity of the TCH.

Upon receiving the TCH assignment message, the base station TCH keys its transmitter and begins sending the SAT. At the same time, the control channel sends an assignment order to the mobile station that contains the frequency and SAT of the TCH that will bear the call.

The mobile station receives the assignment order and retunes its synthesizer and monitors for the designated SAT. Upon SAT confirmation, the mobile station keys its transmitter and regenerates SAT to the base station.

The base station detects the regeneration of the SAT by the mobile station and sends an alert order to the mobile station to cause the mobile station to "ring." The mobile station confirms the alert status by continuously gating the signaling tone into the reverse TCH signal. At this time, and after detecting the presence of the signaling tone from the mobile station, the base station may send a message to the MSC indicating that the mobile station has successfully arrived on the TCH and is alerting.

When the user goes off hook, the mobile station removes the signaling tone from the reverse TCH signal. The base station detects the absence of the signaling tone and sends a message to the MSC indicating that the mobile station has gone off-hook. Conversation may now take place, since both the base station and mobile station have reached the same state.

2.2.5.3 Handoff

Base station 1 detects that the reverse TCH signal strength justifies the consideration of handoff. It sends a handoff request to the MSC containing information relevant to the mobile station (power class, current power level) and its current reverse TCH signal strength.

The MSC receives the handoff request and determines which cells are adjacent to base station 1. In this example, base station 2 is adjacent. Thus, the MSC sends a handoff measurement request to base station 2, as indicated in Figure 2.8.[12]

Base station 2 receives the handoff measurement request and uses its locating receiver to determine the suitability of a handoff. The locating receiver will tune to the reverse TCH frequency the mobile station is currently using, make signal strength measurements on all (or some) antennas, and determine whether or not the SAT color code the mobile station is using can be detected. If the measurements exceed a certain

---

[12]In actual practice, many base stations may be queried in this fashion. Usually, a list is kept at the MSC on a per base station or per antenna basis which shows possible targets of handoffs being sourced by that base station or antenna.

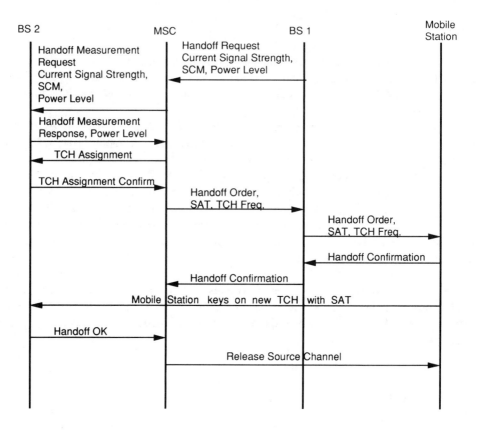

Figure 2.8 Handoff procedure.

criteria for handoff,[13] then base station 2 will send a handoff measurement response to the MSC.

The MSC receives the handoff measurement response and chooses a TCH for base station 2 to use to support the call. The MSC then sends the TCH assignment message to base station 2. Base station 2 receives the TCH assignment message, keys the TCH transmitter, and sends the TCH assignment confirmation message to the MSC.

The MSC receives the TCH assignment confirmation message from base station 2 and sends the handoff order message to base station 1 containing the frequency and SAT of the TCH in base station 2. Base station 1 transmits the handoff order to the mobile station over the forward TCH. The mobile hears the handoff order, confirms the order by gating 50 ms of signaling tone into the reverse TCH signal, and reprograms its frequency

---

[13]Information in the cell site database is usually applied to determine whether or not a response to the handoff measurement request is warranted. The manufacturer may employ sophisticated algorithms to detect the need for handoff and to detect the probability that the new cell will be able to sustain the call.

synthesizer for the TCH in base station 2. The mobile station then looks for base station 2's TCH SAT and, having found the SAT, regenerates it on the reverse TCH to base station 2.

Base station 2 detects the regeneration of the SAT on the reverse TCH and sends a handoff OK message to the MSC. The MSC then sends a release source channel message to base station 1, which causes the forward TCH to be dekeyed, making the channel available for use by another subscriber.

2.3 NAMPS

Narrowband AMPS (NAMPS) is the name given to a more recent air interface compatibility specification recently sanctioned by the TIA. The current interim specification is divided into three parts: IS-88 (air interface), IS-89 (base station requirements), and IS-90 (mobile station requirements). NAMPS is a direct descendent of Narrowband Total Access Communication System (NTACS), which is now enormously popular in Japan.

NAMPS takes each 30-kHz AMPS channel and splits it into three 10-kHz channels. Each cellular call is allocated a 10-kHz channel in NAMPS just as a call is allocated a 30-kHz channel in AMPS. The resulting three-for-one split results in an increase in system capacity without the overhead of cell splitting and all its attendant headaches. NAMPS is compatible with the AMPS system in that the 30-kHz control channel is still used and mobile stations can be built to handle both standards.[14] NAMPS has additional features beyond increased capacity which makes it attractive to service providers.

- Dual-mode operation: The NAMPS standard actually specifies operation in both AMPS and NAMPS channels. Thus, an NAMPS-compatible mobile station may be directed to an AMPS channel, depending on resources available at the dual-mode cell site.
- ARQ signaling: The 200-bps signaling does away with the 10-kHz tones that AMPS uses. It is now possible for the mobile station to acknowledge orders it has received from the base station. This is especially important in the area of handoff, where the AMPS handoff order confirmation of 50 ms of 10-kHz signaling tone was often missed due to interference or incorrect calibration of the base station tone detectors.
- Improved call control: The NAMPS specification provides for a feature known as *mobile reported interference* (MRI). The base station can request that the mobile station send in a measure of the forward TCH signal strength, as well as a measure of the number of errors in the 200-bps signaling stream. The base station may then use this information as further input to the handoff and power control detection software.

---

[14]Note that handoff from an AMPS channel to an NAMPS channel, or vice versa, is possible while using a dual-mode mobile station.

- Short-message service: Alphanumeric messages of 14 characters or less may be sent on the forward channel in a point-to-point or point-to-multipoint mode, thus combining paging functions with cellular service.
- Applicability to portable service: NAMPS requires no significant change from AMPS in terms of technology or packaging. Small, cheap, and lightweight NAMPS portables will be available from the first day of service. This contrasts with the expectation of some commentators that TDMA systems will not achieve small-portable penetration until device and packaging technology catches up with the air interface sometime later in the decade.
- Preloading: Since mobile stations are dual-mode, subscriber equipment to support NAMPS may be sold to the general public well in advance of NAMPS service actually becoming available. Thus, service providers are assured of high utilization of NAMPS equipment when they decide to deploy it.

2.3.1 NAMPS Signaling Specifics

To maintain the 10-kHz channel spectrum usage requirements, a new subaudible signaling scheme is employed between the base station and mobile station. A 200-bps continuous data stream is sent between the base station and mobile station. The user does not hear the signaling, because the frequency resides well below the usable portion of the audio spectrum.

Taking the place of the SAT and signaling tone is a sequence of digital words in the 200-bps stream. NAMPS provides for seven digital SAT (DSAT) sequences that have cross-correlation properties chosen to enhance discrimination between the sequences. Digital signaling tone (DST) consists of the logical inverses of the seven DSAT sequences. The DSAT and DST are used in place of the SAT and signaling tone, respectively, in the same instances as AMPS.

DTX in NAMPS works in a similar fashion to that of AMPS, with the major difference being that the mobile station always scales back its power by three power levels (but not lower than power level 7) while in the DTX-low state. While in this state, the mobile station is required to transmit DST with increased deviation to preserve reverse TCH continuity with the base station.

2.4 OPERATIONAL EXPERIENCE IN THE U.S.

2.4.1 Competition and Subscriber Growth

With the impending breakup of the AT&T conglomerate, the FCC ruled that two construction permits would be issued within each cellular market to promote competition for the good of the consumer. In effect, the FCC created an entire economy. The industry's cumulative capital investment had climbed to $9.3 billion by year-end 1992 [6,7]. Figure

2.9 shows the revenues of service providers. In addition, bear in mind that large cellular infrastructure equipment providers will sell $1 billion of products and services to the cellular market place each year.

Churn, or the percentage of a subscriber base that discontinues service over a period of time, is a major concern of the cellular industry. There are two types of churn: intersystem churn is the percentage of customers that discontinues service with one provider and continues with another within the same cellular market; cellular churn represents those subscribers that terminate their cellular service and do not reconnect with another service provider. It is estimated that monthly intersystem churn is approximately 1.37% and monthly cellular churn is approximately 1.50% [8].

Subscriber growth has outpaced cell site construction growth [6]. Acceptance of cellular radiotelephone as a viable communications medium by the North American public, as well as the profit motive of the service providers, has fueled an astonishing growth rate. Figure 2.10 shows subscriber growth in the United States.

2.4.2 Commercial Issues

2.4.2.1 Service Structure

Each cellular service provider in the top 20 U.S. markets strives to maximize equipment usage and minimize customer irritation. With those ends in mind, a range of service options is offered to the end user. In general, these service options penalize the lower tier

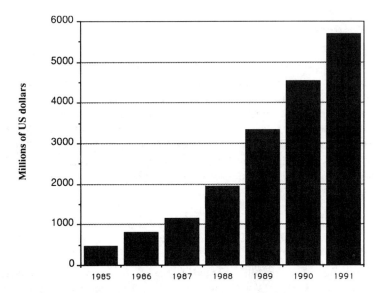

Figure 2.9 Cellular service–provider revenues in the U.S., 1985–81[9].

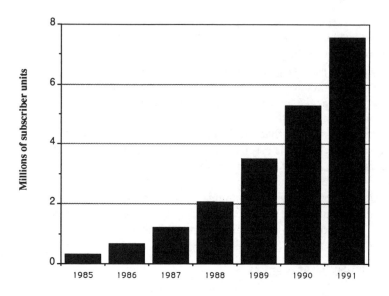

Figure 2.10 U.S. subscriber growth, 1985–91 [9].

users from using their cellular telephones during peak business hours and encourage all users to place calls during off-peak hours.

2.4.2.2 Service Charges

Table 2.3 shows recent service charges for the top five U.S. markets and a function of two commonly offered call plans [10]. Marketing experts have applied their expertise to the problem of baiting customers into using cellular service by using exotic names for their plans like "frequent caller," "corporate," "executive," "economy," and "ultimate." Note that an access charge or monthly service charge is used in addition to charging air time consumption. Some plans lump a certain amount of "free" calling time into their access charges. Packages are also created that lump together several call features (e.g., call waiting, no-answer transfer, etc.) in an attempt to entice further usage of the system by subscribers.

2.4.2.3 Typical Monthly Bill and Cost of Equipment

Table 2.4 shows the recent trend is that subscriber average monthly bills seem to have been falling over the period of 1987 to 1991 [11]. The reasons for this are unclear. Possibilities include increased competition between service providers in each market,

Table 2.3
Examples of Top Five U.S. Market Charges

Market	Plan	Monthly ($)	Peak Hour ($)	Off-Peak ($)
New York	FC	55	0.45	0.45
	OP	25	0.90	0.30
Los Angeles	FC	239*	0.45	0.27
	OP	25	0.90	0.20
Chicago	FC	45	0.32	0.18
	OP	13	0.55	0.13
Philadelphia	FC	155*	0.45	0.27
	OP	11	0.65	0.15
Detroit	FC	50*	0.33	0.16
	OP	22	0.60	0.15

Note: FC = frequent caller, OP = off-peak
* Includes the same amount of call time; 550 minutes.

Table 2.4
Monthly Subscriber Bill

Date	Bill ($)
December 1987	96.83
June 1988	95.00
December 1988	98.02
June 1989	85.52
December 1989	89.30
June 1990	83.94
December 1990	80.90
June 1991	74.56
December 1991	72.74

increased operating efficiency (with savings being passed along to the subscribers), and some profitability being achieved.

Similarly, subscriber equipment prices have steadily decreased due to the aforementioned competition between manufacturers, as shown in Figure 2.11. At some point, this trend will cease as digital technologies come on line and the market for analog tends towards unit replacement. Ultimately, the service providers will have to phase out AMPS if TDMA or CDMA systems deliver on promises of reducing recurring maintenance costs and increasing user capacity.

2.4.2.4 Third-Party Market

Commensurate with the blooming of cellular radiotelephone as a viable communications medium came the emergence of whole support industries. Employment as a direct or

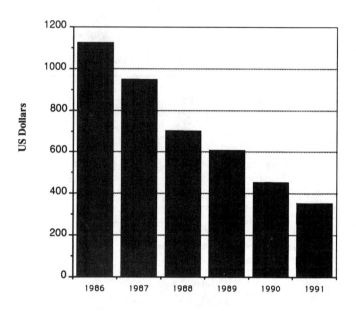

Figure 2.11 Average U.S. prices (wholesale) for installed mobile stations, 1986–91 [6].

indirect consequence of AMPS grew somewhat proportionately to the subscriber popula-tion. Two types of third-party service providers arose. The first type catered to the cellular service provider. Test equipment for cell-site technicians became a small industry and many large and small companies rushed to fill the demand for *cellular service monitors.*[15] As systems grew in complexity and capacity, coverage-quality and frequency-reuse-planning businesses either provided technical expertise or sold equipment to facilitate easier system ''retunes'' and cell planning. Other businesses developed billing and user database software.

The second type of service provider catered to the end user: the subscriber. Businesses that sold and serviced subscriber equipment came into being. These same businesses also added the installation of subscriber equipment in automobiles, limousines, and trucks to their business portfolios. Various retailers of general merchandise and electronics outlets added cellular telephones to their merchandising lines and also resold cellular service within their particular towns and cities. Major hotels and rental car companies offered rental programs for portable cellular phones with daily or weekly rates. Voice mail became an option offered by many service providers. The manufacturing, installation, and maintenance of this equipment also contributed to the trickle-down effect of AMPS.

---

[15]Some of these devices are quite sophisticated and can allow the user to track calls. They will measure signal strength and quality, as well as display the contents of digital messaging between the mobile station and base station. Other devices of this type are used strictly for cell-site RF and audio calibration.

2.5 THE FUTURE OF AMPS

In the near term, AMPS will remain a viable cellular technology, offering enhancement features of interswitch or intermanufacturer handoff (IS-41) and roaming. However, limitations in capacity and over-the-air call features (e.g., ISDN) requiring greater bandwidth will hamper AMPS from competing with future digital technologies.

2.5.1 AMPS Itself

In the United States today, AMPS faces competition from TDMA and CDMA and a host of personal communications systems proposals.[16] In other countries, especially the Third World, AMPS should fare better as more nations are able to afford analog equipment. It is possible that U.S.-based service providers might be able to resell equipment in other nations as part of service-providing partnerships with local PTTs. It will be difficult for the infrastructure suppliers to keep their AMPS assembly lines going once TDMA and CDMA demand starts to soar. The future will no doubt witness manufacturing license agreements in Third-World nations for AMPS. Mobile station manufacturers should fare better, since the replacement market will continue to drive the requirements of keeping assembly lines open and reducing costs.

2.5.2 AMPS in Other Countries

As of May 1992, AMPS is being used in several countries around the world [11]:

Argentina	Australia	Bahamas
Bermuda	Bolivia	Brazil
Brunei	Canada	Cayman Islands
Chile	Costa Rica	Dominican Republic
El Salvador	Guatemala	Hong Kong
Indonesia	Israel	Mexico
New Zealand	Pakistan	Peru
Philippines	Samoa	Singapore
Taiwan	Thailand	United States
Venezuela	Zaire	

At the current time, 90% of all AMPS subscribers reside in the United States. This is probably due to the early introduction of AMPS in that country as well as economic

---

[16]A proposal for a wideband direct sequence CDMA specification is currently in the balloting stage. It is possible that the specification will be complete before the end of 1993 and that manufacturers will begin offering trial systems.

reasons. Indeed, 70% of all subscriber sales were for business use in the United States in 1989. Interest in AMPS is being expressed by the new nation states of eastern Europe (particularly Hungary and Russia). Other countries, like Sri Lanka, have expressed interest in AMPS as a replacement vehicle for, or introduction to, wireline telephone service for their rural populations. While NAMPS is just now getting underway in the United States, several other nations have also shown interest.

REFERENCES

[1] Young, W. R., "Advanced Mobile Phone Service: Introduction, Background, and Objectives," *The Bell System Technical Journal,* Vol. 58, 1979, pp. 1–14.
[2] Electronic Industries Association, "EIA/TIA-553 Mobile Station—Land Station Compatibility Specification," Telecommunications Industry Association, 1989.
[3] Boucher, J. R., *The Cellular Radio Handbook: A Reference for Cellular System Operation,* Quantum Publishing, 1990.
[4] Bell System Technical Reference, Bell System Domestic Public Land Mobile Radio Services, "Interface Specifications for Customer-Provided Mobile Terminals," Pub. 43301, American Telephone & Telegraph, 1982.
[5] Burke, M., and F. Miller, "Fault Tolerance as Applied to Cellular Phone Systems," *IEEE Fault Tolerant Computer Systems,* 21 June 1991.
[6] Herschel Shostek Associates, Ltd., *The Retail Market of Cellular Telephones,* Vol. 9, No. 1, Figure 7.2.
[7] Cellular Telecommunications Industry Association, "Cellular Spurts 17.7% During First Half of 1992," CTIA Press Release, 8 Sept. 1992.
[8] Economic and Management Consultants International, Inc., U.S. Cellular Marketplace, 1991.
[9] CTIA Data Survey, Cellular Telecommunications Industry Association, Washington, D.C., semiannually.
[10] Motorola Cellular Service, Inc., "Rate Plans," 1992.
[11] Steward, S. P., "Continued Growth" and "The World Report '92"; Hinkle, J., "The Top 20"; Bankhead, R., "Eureka! There's Gold In Them Airwaves"; *Cellular Business,* Intertec Publishing, 1992.
[12] *Quantum's Cellular Communication Almanac,* Quantum Publishing, 1990.
[13] Kay, J., and M. Kotzin, "Cellular Systems Technology: Narrow Band Development and Digitally Enhanced Cellular Services," *Pacific Telecommunications Conference Proceedings,* 1992.

Chapter 3
NMT: The Nordic Solution

D. Westin
Ericsson Radio Systems

The Nordic Mobile Telephone System was jointly specified in the late 1970s by the telecommunications administrations of Denmark, Finland, Norway, and Sweden in order to establish a compatible automatic public mobile telephone system in the Nordic countries.

NMT has been fully proven in commercial service in several countries since 1981 and was the first widely spread cellular automatic mobile telephone system. Due to the success of the 450-MHz version of NMT, an expansion based on the same system, but for the 900-MHz frequency band, was introduced when further expansion at 450 MHz became impossible. The first NMT 900 system has been in operation since 1986 in the Nordic countries, and it works in parallel with NMT 450.

3.1 INTRODUCTION

Millions of subscribers on five continents make calls on the NMT cellular telephone systems, which are installed in more than 24 countries. The number of subscribers is increasing still, which is a clear indication of the growing need for mobile communication. This success is a direct result of satisfying system and customer requirements.

The high level of speech quality and service, comparable to that in the public network, is an important factor in the success of the NMT systems. The NMT systems offer basically the same facilities as those offered in the most modern fixed network, and the mobile telephone unit is handled in much the same way as a fixed one.

A considerable amount of work has been done to make the NMT systems as flexible as possible. Each operator can put together precisely the system required with respect to the distribution of channels and functions. The systems also flexibly adapt to the subscriber densities, radio transmission characteristics, and site availability of each system.

In designing the NMT system concepts, effort has been concentrated primarily in the following areas.

- High speech quality;
- High traffic capacity to cope with rapid subscriber growth;
- Upgradable functionality;
- High level of subscriber features;
- Integrated centralized operation and maintenance;
- Low life-cycle cost; and
- High profitability due to short payback period.

The ability to combine the 450- and 900-MHz systems using the same exchange and cell sites has been introduced by most manufacturers. Many operators starting up with low-capacity requirements and/or large areas to cover find it reassuring to use NMT 450, knowing they have the option to increase capacity when they need it.

The experience gained from cellular systems in current operation indicates that market expansion is often limited by traffic capacity. The frequency spectrum is a limited natural resource, but through development of new functionality it has been possible to utilize the spectrum more efficiently.

3.2 MAIN CHARACTERISTICS OF THE NMT SYSTEMS

Table 3.1 shows the main differences and similarities between the two NMT systems.

3.3 SYSTEM OVERVIEW

The Public Land Mobile Network (PLMN) consists of mobile telephone exchanges (MTX), base stations, and mobile stations, as shown in Figure 3.1. The network is modular in design and can be adapted to various capacity requirements by adding further MTXs, base stations, radio channels, and transmission equipment.

3.3.1 System Network

The system is made up of three basic parts: the MTX, the base station, and the mobile station.

The MTX is the controlling part of the system and constitutes the interface to the PSTN. This interface can be made at local, transit, or international gateway levels, but is preferably made at the transit level.

The function of the base stations connected to the MTX over permanent circuits is to handle radio communication with the mobile station. The base station also supervises the quality of radio connections by means of a supervision tone. A base station area or a cell is the geographic area where a call can be effected using one and the same radio

Table 3.1
Comparison of NMT 450 and NMT 900

	NMT 450	*NMT 900*
Frequency band	420–490 MHz	890–960 MHz
Mobile station transmitter	453–457.5 MHz	890–915 MHz
Mobile station receiver	463–467.5 MHz	935–960 MHz
Frequency separation	25 (20) kHz	25 kHz*
Duplex separation	10 MHz	45 MHz
Number of channels	180 (225)	1,000†
Coverage range (base station)	15–40 km	2–20 km
Transmitter output power		
Base station		
Maximum	50W	25W
Mobile station		
High	15.0W	6.0W
Medium	1.5W	1.0W
Low	0.15W	0.1W
Hand-Portable		
High	1.0W	1.0W
Low	0.1W	0.1W
Interface to the PLMN	4-wire transit level	4-wire transit level
Exchange	Typically digital, stored program control	Typically digital, stored program control
Exchange capacity	Up to 1,024 base stations and 65,000 subscribers	Up to 1,024 base stations and 65,000 subscribers
Radio channel capacity	Depends on number of channels, frequency reuse, and cell structure; up to 8,000 channels per MTX	Depends on number of channels, frequency reuse, and cell structure; up to 8,000 channels per MTX
Roaming	Automatic	Automatic
Handover	Intracell	Intracell
	Intercell	Intercell
	Interexchange	Interexchange
Signaling System		
Radio path	FFSK 1,200 bps, compelled signaling, error correction code	FFSK 1,200 bps, compelled signaling, error correction code
To PSTN	CCITT No. 7 signaling and/or MFC R2 signaling	CCITT No. 7 signaling and/or MFC R2 signaling
Roaming	CCITT No. 7 mobile user	Part signaling or MFC-MTX signaling
Interexchange handover	CCITT No. 7 handover	User part signaling
Channel utilization	Flexible handling of calling and traffic channels	Flexible handling of calling and traffic channels
Subscriber facilities	Roughly the same as for subscribers in a modern PSTN	Roughly the same as for subscribers in a modern PSTN
Numbering plan	Country code plus area code plus 6- or 7-digit subscriber number	Country code plus area code plus 6- or 7-digit subscriber number

*12.5 kHz interleaved.   †1,999 interleaved.

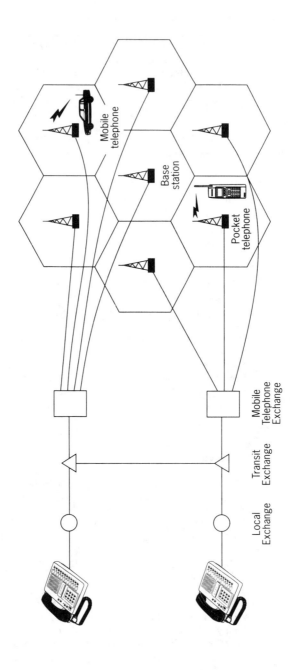

Figure 3.1 System network.

channel. The base stations connected to the same MTX form an MTX service area. An MTX service area can be divided into subareas called traffic areas (TA).

When a call is to be set up from the exchange to the mobile subscriber, a page is sent out in parallel from all base stations in the traffic area in which the mobile station is located instead of being sent out on all base stations in the service area. By doing this, the call setup time and load on the system are decreased.

The mobile subscriber's equipment is called a mobile station, providing a number of facilities for the mobile subscribers. The mobile station may be a vehicle-mounted, transportable, or hand-portable telephone unit compatible with the standard air interface specification used.

The coverage range of each base station is 15 to 40 km for NMT 450 and 2 to 20 km for NMT 900, depending on mast height and actual environment.

3.3.2 Call Handling Functions

All subscribers are able to receive and originate calls in their home MTX as well as in visited MTX areas. When a subscriber moves from one base station to an adjacent base station during a conversation, a handover will take place, enabling the call to continue.

3.3.2.1 Call to a Mobile Subscriber

The call can be routed by the PSTN to the PLMN according to two different principles. The first is that all calls to mobile stations are routed through the PSTN to the home MTX. The second is that all calls are routed the shortest possible way from the calling party to the PLMN. The latter principle means that an MTX, which may be any MTX of the PLMN, will act as a gateway MTX. See Figure 3.1.

The basic method of routing calls to mobile subscribers in NMT is to route the call to the subscriber's home exchange, where information on the current location of the called mobile subscriber is stored. If the called subscriber is located in another MTX area, the call will be routed automatically to the visited MTX. The information regarding the subscriber's location is stored in the subscriber register. The called number (B-number) is received and analyzed in the visited MTX, so that possible active subscriber services, barrings, and so on are detected.

In order to reduce transmission costs, a special routing principle can be used where the call is routed to the nearest MTX. This gateway MTX interrogates the home MTX to find out the location of the dialed mobile telephone. The interrogation request, which is made on a CCITT No. 7 signaling link, enables the call to be set up directly from the gateway MTX to the visited MTX. The interrogation procedure significantly reduces transmission cost, since the calls are set up over the shortest distance through the network. The use of interrogation in a cellular network requires a numbering plan, with which the mobile subscriber can be identified at an early stage of the call setup.

The mobile station number is used at call setup, when the page is broadcast to all base stations within the traffic area concerned. When the mobile station called has acknowledged the call on the calling channel, the exchange will seize a traffic channel and order the mobile station to tune to the selected traffic channel. In the case of no call acknowledgment, a corresponding end-of-selection code is generated and the calling subscriber receives a tone or announcement indicating that the called subscriber cannot be reached for the moment.

If no traffic channel is available, NMT 900 has a function for channel scanning. The mobile station will then start to scan independently for a free channel. A seizure will be sent from the mobile station if a free channel is found, and the call setup is then continued by an identity request. If no scanning order was given, or if a free channel is not found during the scanning time, an appropriate end-of-selection code is generated.

When the mobile station has tuned to the traffic channel, the exchange orders the base station to start transmission quality supervision (see Section 3.3.2.4). The mobile station replies to the identity request on the traffic channel by sending its identity and password (NMT 900). The identity and password are analyzed in the exchange. If the identity is received correctly and the password is correct, the call setup is continued. The location of the mobile station is updated when needed. If the password is faulty, a congestion signal is sent back to the calling party (A-subscriber) and the seized traffic channel is released.

If everything is in order, the group switch is through-connected and a ringing signal order is sent to the mobile station, which generates the ringing signal to the mobile subscriber. The call is set up when the mobile subscriber goes off-hook. The mobile station and the traffic channel are released immediately when the mobile subscriber or the other subscriber clears. After release, the mobile station searches for a calling channel again and locks itself onto it.

In case the call is not acknowledged by the mobile station, the following actions are possible:

- Sending of a recorded announcement to the calling subscriber;
- Sending of a clearing signal to the calling party;
- Call diversion to another number.

When there is no traffic channel available at the base station and the mobile station has acknowledged the call, the call can be set up on a *combined calling and traffic channel*. The combined calling and traffic channel is a feature that is specified for each base station.

3.3.2.2 Call From a Mobile Subscriber

A call from a mobile subscriber starts when the subscriber first dials the desired B-number and then presses the send button. The mobile station then starts to hunt for a free traffic channel or, in NMT 900, an access channel. The mobile station identity and the dialed B-number are automatically transmitted to the MTX. The MTX checks the calling subscrib-

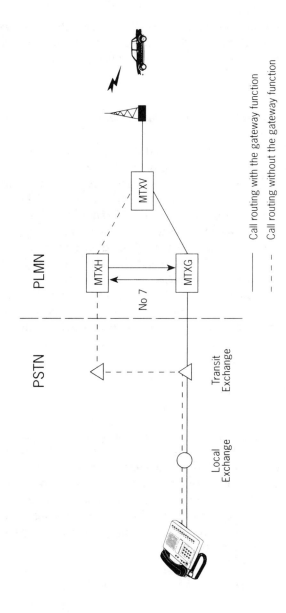

Figure 3.2 Call routing with and without a gateway function.

er's categories and, in NMT 900, the password to verify that the caller is permitted access to the system. If the check is successful, the call setup is continued. When needed, the location of the mobile station is updated.

The dialed number is also analyzed to determine if the caller is allowed to call that number. If all analyses are affirmative, the call is set up and the supervision of speech quality is started. The release is done in the same way as in the case of a call to the mobile subscriber.

3.3.2.3 Access Channel for Seizure From Mobile Station

The purpose of this function is to speed up and enhance the setup of calls from mobile subscribers. The function is only available in the NMT 900 system.

An access channel can be defined for use instead of free-marked traffic channels for seizures from mobile stations. The call setup is made more quickly than via traffic channels due to the subbanding of the access channel band from which the mobile station searches for a channel to make an access. An idle traffic channel is selected and started after reception of seizure on the access channel. After this, a channel allocation order is sent to the mobile station and the call setup continues on the allocated traffic channel.

Traffic channels are located outside the access and calling channel bands. The access channel has a high capacity, and one access channel per base station is therefore sufficient.

The mobile station receives information about the access channel band in the signaling on the calling channel. The access channel is also used for calls when no traffic channels are available.

3.3.2.4 Transmission Quality Supervision

The purpose of this function is to maintain the best possible transmission quality of a call in progress, irrespective of a subscriber's movement within the service area. This is done by choosing the most suitable base station based on the signal-strength measurement results, performed at the current and all the neighboring base stations.

Transmission quality is supervised by the base station in two ways. The signal strength of the carrier from the mobile station is measured. The signal-to-noise ratio (S/N) of a special supervision signal is measured. This supervision signal is sent from the base station and looped back from the mobile station in the traffic channel. This signal (also called the *phi-signal* or *pilot signal*) is a tone above the speech band. Four different frequencies with 30-Hz separation are used.

- Signal number 1: 3955 Hz
- Signal number 2: 3,985 Hz
- Signal number 3: 4,015 Hz
- Signal number 4: 4,045 Hz

When transmission quality drops below a certain limit, the base station informs the exchange. The MTX will then request the signal-strength receiver equipment in the neighboring base stations to report the results of the RF carrier signal strength. The results are evaluated and ranked by the MTX. The consequences may be a power increase or decrease in the mobile station, handover to a new channel in the same or a different base station, or no action at all.

3.3.2.5 Blocking of Disturbed Channels

Idle traffic channels disturbed by in-system or out-system interference are automatically blocked for the duration of the disturbance and are thus not used for traffic, new calls, or handovers.

3.3.2.6 Mobile Station Identity Check

This function is used to prevent unauthorized use of a mobile station and is only available in the NMT 900 system. For instance, a stolen mobile station can be put out of service while the original mobile subscriber can still use the same mobile subscriber number as before. With new mobile station equipment, only a password is changed.

The mobile station password is a three-digit part of the mobile station identity. It is stored in the mobile station and in the exchange, but is not known to the subscriber. The password is signaled at the identification of a mobile station.

The validation of the password can be checked on calls to and from mobile subscribers and also on roaming updating messages. When detection of an incorrect password occurs, the call is disconnected and location or subscriber-controlled category updating is not performed. Supervision of the identity check is available, informing the operator of repeated call attempts to or from mobile stations with illegal passwords.

3.3.2.7 Subscriber Identity Security

NMT subscriber identity security (SIS) is a system that improves the security of the subscriber identity in NMT 900 beyond what is possible with the three-digit password used for mobile station identity check.

The function authenticates the mobile station identities and protects subscribers from illicit use of their identities. The authentication is based on a challenge-response method between the MTX and the mobile station, including encryption of the dialed B-number. A secret authentication key (SAK) is installed in the mobile station. The SAK is also stored in the MTX's authentication register (AR).

The identity is checked every time a call is made from the mobile station. The check is performed by the MTX sending a random number to the mobile station, which computes an answer by using its SAK. The answer is sent to the MTX, which compares the answer to its own computed result, and when they correspond, access is allowed. If correspondence is not achieved, the call is rejected. This method makes it impossible to intercept the secret key.

The MTX is to be able to handle mobile stations with and without SIS. The function is category-controlled. When roaming, the category information is transferred to the visited MTX via the mobile user part (MUP).

3.3.2.8 Discontinuous Reception

This function is also called *battery saving* and it is only available in the NMT 900 system. Its purpose is to save battery energy in the handheld mobile stations. Battery saving is achieved because the receiver in the mobile station is switched off most of the time, with only a clock function active during "sleeping time." Calls received during this time are buffered in the exchange. The paging for buffered calls is sent when sleeping time is over and a new battery-saving order is sent out again to mobile stations after the pagings have been sent.

3.3.3 Radio Aspect

3.3.3.1 Frequency Bands

The connection between a base station and a mobile station consists of a full-duplex radio channel allowing information to be exchanged simultaneously in both directions. The base station transmitters operate in the high-band region and the mobile transmitters in the low-band region.

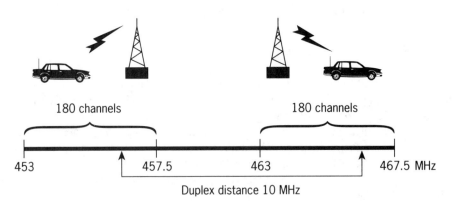

Figure 3.3 Frequency band for NMT 450.

In the 450-MHz range, the duplex distance is 10 MHz. The frequency band offers 180 channels with 25-kHz channel separation. See Figure 3.3.

The channel separation in the NMT 900 system is normally 25 kHz, which corresponds to about 1,000 channels. However, mobile stations and base stations for the NMT 900 system are specified for 12.5-kHz channel separation (interleaved channels). This means that the system can use a total of 1,999 channels. See Figure 3.4.

3.3.3.2 Channels

Communication between a mobile telephone exchange and a mobile station is performed via routes known as *channels*. The channels are divided into groups, referring to the channel functions.

Calling Channel. Each base station uses one channel as the calling channel. On this channel, the base station transmits a continuous identification signal. Mobile stations within the base station area lock onto the calling channel. On a calling channel, calls are transmitted to a mobile subscriber who is expected to be within a given geographic area. After the mobile station has responded to the calling signal, the exchange allocates another channel, a traffic channel, to the mobile station over which the conversation can take place.

Traffic Channel. The traffic channel can have three different conditions: (1) the traffic channel with ''free marking'' is mainly used for setting up calls from mobile stations, (2) a ''busy'' traffic channel is occupied by a call, and (3) an ''idle'' traffic channel represents a traffic channel that is not in use at the moment (i.e., not occupied and not free-marked).

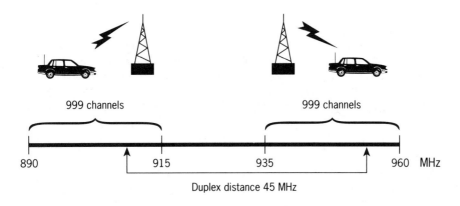

Figure 3.4 Frequency band for NMT 900.

Subbands. In NMT 900, parts of the total frequency band are arranged into narrow subbands. These are the CC band and access channel band. The remainder of the frequency band consists of traffic channels. The purpose of this is to make call setup quicker.

Access Channel—in NMT 900. To increase the capacity in the traffic channel, an access channel can be used instead of the free-marked traffic channel. At call setup, over an access channel, a mobile station will immediately be directed to a free traffic channel outside the access channel band.

Combined Calling and Traffic Channel. The calling channel of a base station can be defined to operate as a combined calling and traffic channel. This implies that if all traffic channels are occupied, the mobile station can use the calling channel to set up a call. The base station will then, for a short period of time, completely lack the calling channel. But as soon as a radio channel becomes free, it functions as a combined calling and traffic channel.

Data Channel. The purpose of the data channel is to make signal-strength measurements on mobile stations in ''conversation state'' on order from the MTX. The result of the measurement is used by the MTX at handover. The data channel is of the same design as a channel unit, but without the transmitter.

At each base station, the MTX reserves the last free channel. This is used for calls to the mobile telephone or for handover from a nearby base station.

Every base station should generally have one calling channel, one access channel or some free traffic channels, and one data channel. The rest of the available radio channels normally have the function of idle traffic channels.

3.3.4 Mobility Management—Multiple MTX Networks

3.3.4.1 Paging

This function is used to determine the location of the mobile station. The service area of an MTX can be divided into a number of traffic areas. Paging means the sending of a call frame with the called mobile station number over all calling channels in the traffic area where the subscriber is expected to be. If the traffic area information is missing for the called subscriber, paging will be performed on all traffic areas in the MTX area concerned. The mobile station will then recognize the call and respond at the particular base station where it is located. If no call acknowledgment is received, the call paging is repeated once.

Paging is considered unsuccessful when a certain period of time is exceeded without call acknowledgment from the mobile station. By paging a mobile subscriber only within the traffic area in which the last registration was made, the paging load on the system is reduced.

3.3.4.2 Handover

Switching a call in progress from one base station to another (handover) is performed to allow the call in progress to continue while the mobile subscriber is moving out of the coverage area of the current base station.

The radio connection quality is measured and evaluated continuously during the call, and if the quality deteriorates, an alarm is sent from the base station to the MTX. The exchange investigates if a better base station can be found by ordering a signal-strength measurement on the surrounding base stations. If a better base station with an available radio channel is found, a handover will be initiated.

If not, the call is continued in the current channel. New signal-strength measurements and handover attempts are made periodically until the handover attempt is successful, or the call must be disconnected due to the fact that the subscriber has moved too far away from the serving base station. Normally 20- to 30-sec periods are used between the attempts.

The handover includes seizure of the most suitable channel in the new base station, starting with the transmission quality supervision of the new channel, and switching of the speech path towards the new channel. The MTX transmits an order to the mobile station to change the frequency to the selected new traffic channel in the new base station. The switch is made in the MTX at the same time as the mobile station changes its frequency. After a successful handover, the old channel is released.

3.3.4.3 Intracell Handover

This function is used in order to hand over mobiles from a disturbed traffic channel to another traffic channel in the same base station, thus improving speech quality.

3.3.4.4 Handover Queue

When the system is highly loaded with traffic, the handover may suffer from channel congestion, the result being that handover cannot be performed.

After carrying out signal-strength measurements on surrounding base stations, the MTX stores the first and the second best alternatives. If a handover is required and the base station with the best value does not have a traffic channel available, the second best base station is chosen. In case this base station does not have a traffic channel available either, the handover will be queued on the best base station for a predetermined time. The maximum queuing time is set separately for each base station and is adjustable between 0 and 10 sec.

When a traffic channel becomes available for a queued handover, the channel shall be reserved and the call switched in the normal way. If no traffic channel becomes available during the queuing time, the corresponding queued handover will be taken away from the queue, the handover attempt will be terminated, and the call continues on the old channel. The queuing does not take place for the second best base station.

3.3.4.5 Location Updating

The purpose of this function is to keep continuous track of the mobile station in the network. The function comprises the automatic updating call from a mobile station and the updating of the location data in the MTXs. The location data indicate the current traffic area where calls to the mobile station can be directed.

An idle mobile station is locked to a calling channel. When the mobile subscriber moves out of reach of this channel, the mobile station searches for a new calling channel to lock itself to. If this calling channel belongs to a different traffic area, the channel indicates this and the mobile station will initiate a location updating call. The location data is then updated in the exchange.

If for some reason the location updating was not successful, an alarm indication is given in the mobile station and manual updating is necessary; that is, the subscriber simply presses the Hands-Free key on the control head or picks up and replaces the handset.

3.3.4.6 Roaming Updating

Mobile subscribers may move freely in the coverage area of the system and still be provided full service. This feature is called *roaming*.

Each mobile subscriber is registered permanently in its home MTX, where all information pertaining to it is stored. When a mobile subscriber is roaming to another MTX area, which is controlled by a visited MTX, roaming updating between MTXs is initiated. This procedure uses end-to-end signaling with a specific mobile user part (MUP). The visited MTX updates the home MTX regarding the subscriber's location, and in return receives a copy of the mobile subscriber's status data.

When the mobile subscriber leaves a visited MTX (say, MTXV1) and moves to another MTX area (say, MTXV2 or the home MTX), it will update itself as described above. However, when it has been updated, the home MTX must order MTXV1 to erase data stored there concerning that mobile subscriber.

3.3.4.7 Interexchange Handover

The interexchange handover function is an extension of the handover function to allow the switching of calls in progress, even to base stations controlled by other MTXs. Together with the automatic roaming function, interexchange handover makes the mobile subscriber independent of the borders between MTXs. The signaling required, as in the case of roaming, is based on CCITT No. 7 and a specific handover user part (HUP).

Since this function deals with more than one exchange, it is necessary to give the exchanges different names. The anchor exchange controls the service area where the mobile station was at the original call setup, the serving exchange has radio contact with the mobile station, and the target exchange is the exchange where the serving exchange has located the most suitable base station for the handover.

There are two main traffic cases: basic handover and subsequent handover. The basic handover is done from an anchor exchange to a target exchange. The subsequent handover is done either from a serving nonanchor exchange back to an anchor exchange, or from a nonanchor serving exchange to a third exchange.

The interexchange handover is controlled by the anchor exchange. Connection between the anchor exchange and target exchange is always direct, even in the case of a third exchange, providing path minimization. See Figure 3.5.

3.3.4.8 Subscription Areas

This function enables the operator to define the allowed mobility of a subscriber. The term *subscription area* means a restricted geographical area where the subscriber is permitted to make and receive calls. This area can be on the base station level up to traffic area level. Two practical applications are:

- For a fixed mobile station in an area where normal telephone wiring would be more expensive. The subscription area could be only one base station.

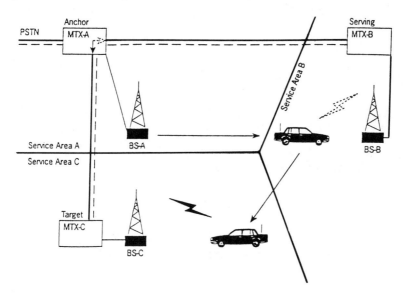

Figure 3.5 Subsequent handover to third MTX.

- For a rural subscriber who can use his or her mobile telephone in the whole system coverage area except for large city areas. A special tariff can be used for this category of subscriber.

CCITT No. 7 MUP signaling is required to implement this function in areas covering more than one MTX.

3.3.5 Cell Planning

3.3.5.1 The Concentric Model

To allow high traffic density in large urban areas, a special small-cell technique has been developed. The method is called the *concentric model* and was first implemented in the NMT 450 network in Stockholm in 1985. The basic idea is to provide a very high capacity in the central part of the city where the traffic density is the highest. The capacity per area unit then gradually decreases outwards from the densest traffic in the center, as indicated in Figure 3.6.

This small-cell model is characterized by the placing of base station sites on concentric circles. All frequency groups can be repeated on each circle, but a certain group must not have the same transmitting direction on other circles. The angle between cells using the same set of frequencies should be about 120 deg. This cell structure requires low output power and antennas with front-to-back attenuation of approximately 25 dB.

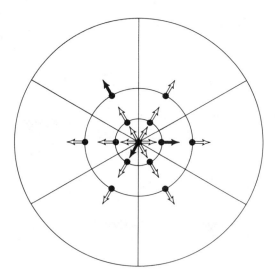

Figure 3.6 The concentric model.

A very important objective with the concentric model is to find the peak traffic density area and place the center site at that point. This area does not necessarily have to be the center of the town.

3.3.5.2 Interleaved Channels in NMT 900

The use of interleaved channels means that the capacity in the system can be very much improved, but the channel assignment will be more complicated. The frequency space between a main channel and its interleaved channel is 12.5 kHz; this means that an interleaved channel could not be assigned to a cell next to another that uses the adjacent main frequency.

A frequency plan that optimizes the reuse of frequencies has to meet the following criteria.

- Each channel group should be repeated only once within a cluster of 19.
- Channel groups with a distance of 12.5-kHz should not be used on cells bordering each other.
- Two channel groups should be available on the same cell.
- Regularity must be maintained so that the same conditions exist in areas located on both sides of the border of two clusters.

Figure 3.7 shows channel assignment in a cluster using both ordinary and interleaved channels. Since each cell uses two channel groups, the capacity increases by more than 100% compared to a system without interleaved channels.

3.3.5.3 Large-Coverage Base Station

A large-coverage base station is a base station with a radio coverage area large enough to overlap the area of several normal base stations. A large-coverage base station is sometimes also called an *umbrella* base station.

The large-coverage base station may be used to cover an area which is difficult to cover with small-cell base stations. This also means that higher transmission powers are used than in small-cell base stations. To minimize traffic on a large-coverage base station due to poor channel utilization efficiency, traffic is connected via channels on such a base station only if no channels are available on ordinary base stations. In addition, traffic is handed over to an ordinary base station immediately after a channel becomes available.

3.4 NMT SYSTEM FEATURES

3.4.1 Subscriber Services

The general philosophy of NMT is to offer the mobile subscriber the same services as those offered to subscribers in the fixed network. The mobility of the subscriber demands

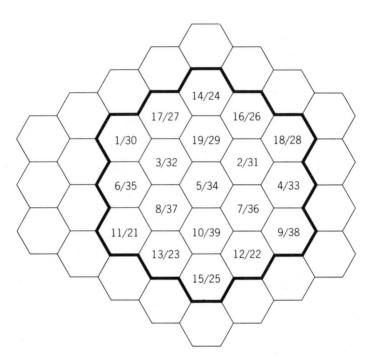

Figure 3.7 Use of interleaved channels.

sophisticated solutions in order to maintain these services unchanged throughout the network. Information about the individual subscriber access to and current status of a specific subscriber service is transferred between the MTXs as the subscriber roams. Innovative services provide added value for subscribers in the mobile network.

3.4.1.1 Subscriber Services Common for Fixed and Mobile Networks

The central part of the system, the MTX, provides mobile subscribers with basically the same services as those of the PSTN. It also provides the possibility of incorporating new services as they are developed for the fixed network.

Table 3.2 presents the subscriber services that can be accessed by the subscribers in the PLMN as well as the PSTN.

The malicious-call tracing service enables the origin of malicious, nuisance, and obscene calls to be traced. In general, there are two ways of achieving this. The first way is to hold the call and use manual call tracing. Registration of information, including the calling subscriber's number, will be performed manually. The second way is an automatic indication (printout) of all calls to the subscriber who has the malicious-call tracing service

Table 3.2
Subscriber Services

Call Barring	Inquiry	Absent Subscriber and Interception
Call diversion	Hold for inquiry	Do not disturb
Immediate call diversion	Hold for inquiry with transfer	Call waiting
Call diversion on no reply or no acknowledgment	Add-on conference	Malicious-call tracing
Call diversion on busy or base station congestion		

Note: Subscribers can also generally deactivate services.

activated. Malicious-call tracing in NMT is done using the latter method. A prerequisite for malicious-call tracing is that the A-number is transferred from the PSTN to the MTX.

3.4.1.2 Subscriber Services Implemented in the MTX

Mobile subscribers may also benefit from services specific to the mobile radio system. The services are available for the mobile subscribers both at the home exchange and at visited exchanges.

Data Mobile Station. This service implies that, for mobiles equipped for data transmission, all signaling (e.g., for signal-strength measurements and handover) is prevented on the radio channel in order to not disturb the data transmission. The data transmission service can be activated or deactivated by the subscriber.

Coin Box Mobile Station. Subscribers having the coin box mobile station category will receive charging information to the mobile station during calls from the mobile. The mobile station may either be a money-operated mobile telephone or a standard mobile station that can display the charging information.

DTMF Signaling. DTMF signaling from the subscriber's equipment is used to control various types of automatic devices such as telephone answering machines, voice mail systems, and so on. The special DTMF signaling specified by NMT ensures that the transmission of DTMF digits is error-free, even when interference occurs on the radio path.

Priority Call. This service will give subscribers with a specific priority category an enhanced capability to access the system. They can initiate calls on the calling channel as well as the traffic channels. It is also possible to reserve a number of channels for subscribers with priority on base stations assigned an emergency status. For priority calls, a priority mobile is needed.

3.4.1.3 Subscriber Services Implemented in the Mobile Station

The market offers a range of mobile telephones and hand-portables complying with the NMT standard. Although the functional content of features implemented in the telephones varies among different brands and models, the following features are implemented in most mobile stations.

- Push-button dialing;
- On-hook dialing;
- Abbreviated dialing, numbers stored in mobile station;
- Last-number redial;
- Electronic lock;
- Service indicator; and
- Roaming alarm.

As options, the following functions are normally available.

- Hands-free operation;
- Modem interface for data communication;
- Rear-seat extension;
- Voice-controlled hands-free; and
- External horn alert.

3.4.2 Charging

3.4.2.1 Charging Principles

The charging principles in the PSTN are normally also used in the PLMN. This means that the call charge is based on origination, originator category, dialed number, and terminating category. Charging of calling party is used for calls to and from mobile stations. Normal PSTN charging methods can therefore be used without modification when the mobile telephone service is introduced. Charging of mobile-originated calls is done by means of toll ticketing (TT). The TT output is performed by the MTX where the call was originated. It is, however, also possible to charge the mobile subscribers for incoming calls. The toll ticket will, in this case, be output from the MTX where the call was terminated.

3.4.2.2 Charging Analysis

Charging analysis provides the basis for the charging of calls and services. One of the items included is the location in the network where the charging of the particular call should take place. Normally, for local traffic, the charging point is the originating exchange. Included in the charging analysis is the tariff class for the call. An example of the tariff class is "calls within own charging area." Different sets of input parameters can result in the same tariff class. A tariff switching program calculates the different tariffs to be used at various times. This implies that each tariff class could include different tariffs, depending on the time and type of day.

3.4.2.3 Charging Data Recording

Charging data recording (CDR) comprises functions for teleticketing (TT). Each individual is assigned to a TT-charged call. In the CDR, all data pertaining to call charging are collected and sent to the TT block for output when the connection is cleared. The data included in the TT records is also sufficient for international accounting.

3.4.2.4 Toll Ticketing

TT comprises functions for the editing and output of TT data. It receives the data from CDR, edits the data to the required format, and sends it to file management subsystem (FMSC) for output. The standard output of TT data is ISO (ASCII) coded. Some examples of information included in the TT record include:

- Calling party subscriber number;
- Called party subscriber number;
- Called party category;
- Date and time charging begins;
- Chargeable duration;
- Incoming and outgoing routes; and
- Tariff class.

The charging subsystem has been specially tailored to meet the unique needs of cellular telephony. For instance, different record formats can be used for incoming and outgoing calls. Indication of abnormal call release can also be included.

3.4.2.5 Output Devices

The standard output devices are magnetic tape units. Data link output, using X.25, is also available. See Figure 3.8.

Billing

Billing is normally made by offline processing of the TT charging data.

3.4.2.6 Accounting

The accounting functions are used to collect and output data needed for financial reconciliation purposes between the PLMN and the PSTN operators. The accounting analysis covers:

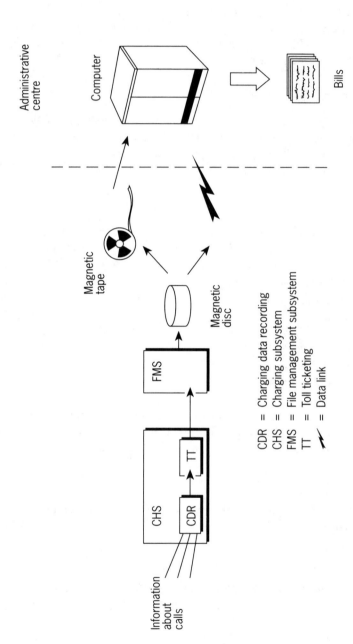

Figure 3.8 Standard output devices.

- Destination;
- Outgoing route;
- Incoming route;
- Calling subscriber category;
- Calling subscriber number, received language/discriminating digit.

If the result of the analysis indicates that accounting is to be performed, an accounting class to be used for the call is assigned.

Accounting Data Collection

Collection of accounting data begins when the called subscriber answers. The collected data are number of calls, call duration, and number of meter pulses sent, all stored in counters according to accounting class.

Output of Accounting Data

The time of day and the interval between output of all or part of the accounting classes may also be ordered and performed directly as an answer to a command. Any standard I/O device can be used.

3.4.3 Traffic Measurement and Statistics

The traffic measurement and statistics features play a very important part in the management of the cellular network. The extremely high subscriber growth demands sophisticated tools for network planning. Some of the most important measurement and statistics functions in NMT are briefly described below.

3.4.3.1 Cell Traffic Statistics

The Mobile Telephone Cell Traffic Statistics function is used to collect traffic data in the cellular system. The function makes it possible for system operators to get a picture of the traffic in the system, both regular and irregular.

The number of certain cellular- or telephony-related events during a defined period of time is collected by means of counters and output to an I/O device. The events recorded are directly associated with the signaling between mobile station, base station, and the exchange.

The event results can be collected for the whole exchange or for specified base stations. There is also a possibility of collecting data on handover events between a base station and its neighbors.

Cell Traffic Recording

This function records and outputs detailed information concerning traffic events in the cellular network. Cell Traffic Recording implies the recording of traffic events and relevant data during a call to or from a mobile subscriber. Registration is performed on a per-call basis. Recording can be restricted to certain mobile stations or traffic channels. Call-related and/or handover-related events can be selected. The events to be recorded are selected by specifying a measuring program.

3.4.3.2 Traffic Measurement on Base Station Routes

This function includes the measurement of traffic through the base stations in the network. The principle of traffic measurement is to measure only real traffic on the traffic routes and only data in the data routes. Therefore, no registration is made for signaling on the calling channel, signal-strength measurements are only registered on the data route, and no registration is made for the automatic or manual test blocking of channel devices.

The following data are calculated for each route in a specified measurement period.

- Number of devices;
- Number of blocked devices;
- Number of call attempts;
- Call congestion;
- Traffic intensity;
- Mean holding time.

The *number of devices* gives the number of traffic or data channels connected to the traffic or data route during the measurement period.

A *blocked device* is removed from traffic by all types of automatic or manual blocking, except test blocking.

A *call attempt* is registered when the exchange receives an accepted seizure, an accepted updating seizure, or an accepted-call acknowledgment from a mobile station. Registration is also made when there is a seizure of a channel for handover or when a signal-strength measurement order is sent to the base station. No call attempts are registered for incoming data routes.

Call congestion is registered when no traffic channel is available after a call acknowledgment or when no signal-strength measurement order can be sent within a specified time because the traffic exceeds the data channel or traffic channel capacity. No congestion is registered for incoming traffic or data routes.

The *traffic intensity,* in erlangs, gives the average number of devices seized for the traffic direction in question during the measurement period.

The *mean holding time* provides the average time a call stays in the base station. Automatic updating calls are also included.

3.4.3.3 Processor Load and Exchange Input Load Measurements

This function is used to carry out measurements on processor load and exchange input load. Load data can be obtained by defining a measurement program for scheduled measurements. The processor load is defined as the percentage of time that the processor spends executing tasks above a certain priority level. The exchange input load is defined as the total number of calls offered to the exchange.

3.4.3.4 Traffic Measurement on Routes

The purpose of Traffic Measurement on Routes is to get background information for long-term planning and to check the dimensioning of the routes. Traffic recording is controlled by a preprogrammed time schedule connected to a measuring program. A measuring program is defined by a time schedule, recording groups (containing routes to be measured), printout formats, and output device. Several measuring programs can be used at the same time independently of each other.

3.4.3.5 Traffic Measurement on Traffic Types

The Traffic Measurement on Traffic Types function is used to record data on the different traffic types existing in an MTX (e.g., incoming, outgoing, originating, terminating, internal, and transit traffic). The purpose of the function is to provide information for control and followup of the traffic situation in the MTX. Furthermore, the function can be used in connection with long-term traffic planning. The measurements are performed according to a preprogrammed time schedule.

3.4.3.6 Traffic Dispersion Measurement

This function measures certain data on a per destination code basis. The destination code is the part of the called number identifying any exchange or numbering area within the network. The data obtained from this function can be used for network planning purposes; for example, the traffic matrix is formed from traffic dispersion measurements. This matrix provides a basis for forecasting the network traffic, information that is necessary for planning networks employing alternative routing.

3.4.3.7 Data Recording per Call

This function provides data that can be used for many different purposes by sampling a number of calls from a specified object, such as a base station or a particular subsystem. The information obtained from the recorded sample can be used to determine, for example:

- Call dispersion;
- Traffic dispersion;
- Interarrival times of calls;
- Holding time distribution; and
- Conversation time distributions.

The data output is on a per-call basis. It can subsequently be postprocessed to give the actual information wanted. The function is run according to a preset time schedule.

3.4.3.8 Charging Statistics

The Charging Statistics function is used to record data that can serve as a basis for tariff planning and deciding whether and how a tariff change should be performed. It also makes it possible to observe the effects of a change of tariff.

3.4.3.9 Service Quality Statistics

The basic function for judging the overall quality of service is Service Quality Statistics. This function works automatically according to conditions given by means of command, and performs automatic service observation on a random sample of calls fulfilling a preselected traffic criterion. Examples of registered data are:

- Number of calls with a postdialing delay exceeding a preset value;
- Number of calls where conversation time is shorter than a preset value; and
- Number of calls meeting congestion.

3.4.3.10 Traffic Observation

As a complement to the above-mentioned function, the Traffic Observation function can be used. This function provides a listening connection to an operator for a manual classification of parameters concerning the quality of service experienced by the subscriber.

3.5 SIGNALING SYSTEMS

3.5.1 MTX to PSTN

The signaling systems used between an MTX and the PSTN are individually designed or selected for each MTX, depending on the exchanges of the PSTN to which the MTX is connected. This individual signaling system design is based on standard systems, such as CCITT No. 7 telephone user part (TUP), R2, or national signaling systems.

3.5.2 MTX to MTX

MFC and/or CCITT No. 7 common channel signaling is used between MTXs for telephony signaling. The advantages of No. 7 compared to MFC signaling are that it can transfer more data, is faster and more cost efficient, and has a higher capacity.

The signaling system No. 7 protocol used in NMT is a CCITT common channel signaling system protocol. Roaming and inter-MTX handover signaling is performed using the MUP and HUP. The transport layers are handled by CCITT No. 7 signaling connection control part (SCCP) and the message transfer part (MTP).

3.5.3 MTX to Mobile Station

Binary signaling is used for the exchange of messages between MTX and mobile stations. The signaling between MTX and mobile station via base station handles the setting up and clearing of calls, handover, updating, and the ordering of change of transmitting power, and network-managing functions. The signaling between MTX and base station transfers the alarms and handles remote control of the base station.

3.5.4 1,200-baud Signaling

The signaling equipment consists of modem and coding equipment (see Figure 3.9). In the MTX there is a function for the equalization of group delay distortion.

3.5.5 Modulation

The binary position in the signaling is transferred in accordance with the FFSK method. With a data rate of 1,200 bps, logical one is represented by one period of the frequency 1,200 Hz and logical zero by 1.5 periods of the frequency 1,800 Hz. See Figure 3.10.

The transition between a one and a zero occurs at the zero-transition points of the signal. This implies that the signal is always correct in phase, and therefore can simply be received as a coherent signal (receiving with known phase).

The transfer of signal frames, of course, is done over the whole defined transmission channel, and thus also between base station and mobile station.

3.5.6 Basic Structure

All signaling is done with the aid of frames, which are always of the same length and consist of 166 bits. The frame structure is shown in Figure 3.11.

During reception, the first 15 bits are used for bit synchronization; that is, the clock pulses in the receiver of the modem are so adjusted in time that every bit is read off at

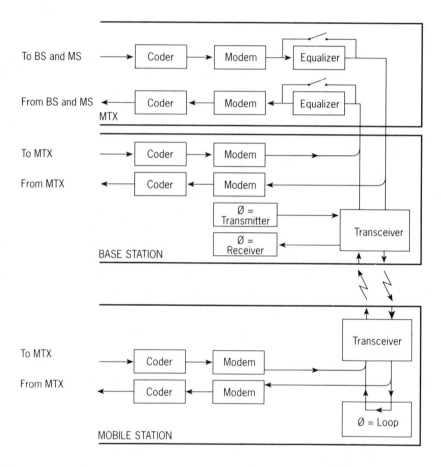

Figure 3.9 Function schematic block diagram of the signaling equipment. The equalizer is only to be found in the MTX.

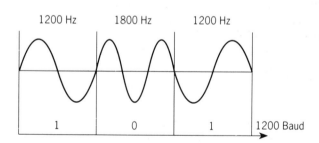

Figure 3.10 FFSK modulation.

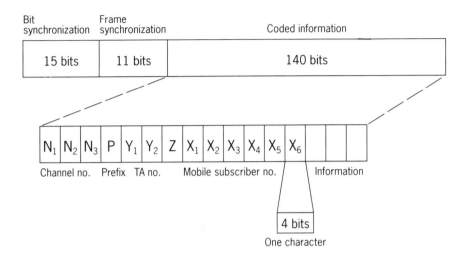

Figure 3.11 The basic structure of a signaling frame.

its middle. The sequence pattern for bit synchronization is alternating ones and zeros (101010101010101). With good signal strength, one sequence is enough to attain bit synchronization.

Then follows an 11-bit sequence, which is used to detect where the information in a frame begins (i.e., frame synchronization); the sequence pattern is 11100010010.

The remaining part contains the information of the frame. This part is, however, coded so that an original information of 64 bits is expanded to the length of 140 bits.

To cope with the fading characteristics on the radio path, an error-correcting code is used. The code used is a Hagelberger convolutional code. This code can correct error bursts of up to 6 bits, provided that the distance between two error bursts is at least 20 bits. The code is able to correct most errors due to fading, which appear at typical mobile driving speeds.

Compelled signaling is used throughout, providing a highly secure system. Of particular interest are the times shown in Table 3.3.

3.6 RADIO BASE STATION

3.6.1 General

The radio base station is connected to the MTX on point-to-point circuits. It handles the radio communication with the mobile stations and functions chiefly as a relay station for data and speech signals. The base station also monitors the quality of the radio transmissions in progress by means of the supervisory signals and by measuring the strength of the signals received from mobile stations.

Table 3.3
System Timings

System	Time (seconds)
Call from mobile setup time	4
Call to mobile	
Time on a calling channel	1
Time on a traffic channel	1
Call clearing time	0.75
Handover time	
NMT 900	0.3
NMT 450	1

The base station can be installed, for example, in an office building in cities or in a container in the countryside. The base station is a set of equipment to serve a number of cells, normally one omnidirectional cell or sectored cells. See Figure 3.12.

The base station comprises the following equipment.

- Channel units;
- Transmitter combiner;
- Receiver multicouplers;
- Signal-strength receiver;
- Reference oscillator;
- RF test loop (RFTL);
- Power supply;
- Power monitoring unit;
- Antenna system.

Each channel unit consists of a transmitter, receiver, control unit, and a power amplifier connected to the transmitter output.

3.6.2 Transmitter

The maximum transmitter output power is: for NMT 450, 50W; for NMT 900, 25W. The output power determines the size of the coverage area of the channel in question. It can be adjusted manually to obtain the required coverage. The transmitter contains a compressor (NMT 900), a phi-signal modulation, and a pre-emphasis.

3.6.3 Receiver

The receiver contains: an expander (NMT 900), diversity (NMT 900), de-emphasis, phi-signal detection, band-stop filter phi-signal, and signal-strength measurement.

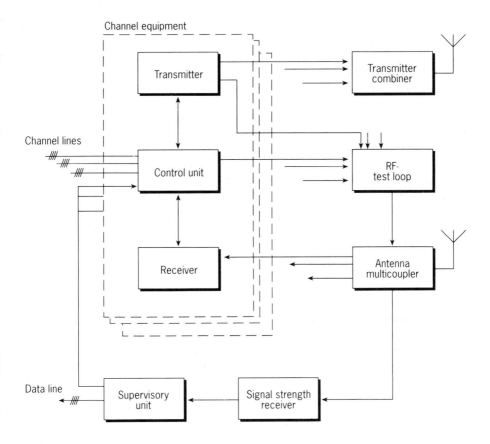

Figure 3.12 Block schematic diagram for the base station.

3.6.4 Control Unit

Each transmitter-receiver pair is managed by a control unit consisting of a microprocessor system, a modem for signaling with MTX, and circuits for generating the phi-signal. The control unit directs special functions of the respective channel equipment and passes signals between the MTX and the radio channel. The functions handled by the control unit are the following.

- Transmitter on and off switching;
- Setting of the channel number;
- Transmission of channel information backwards to the MTX;
- Fault alarms given to the MTX;
- Loop connection of the incoming channel lines for test purposes;
- Check of the RFTL for the channel equipment;

- Generation of the supervisory signal (phi-signal) and evaluation of this signal returned from the mobile station;
- Alarm for poor speech quality;
- Performance of self-test;
- Supervision of transmitter and receiver.

3.6.5 Supervisory Unit

The base station contains a supervisory unit for signal-strength measurement that is common for the base station (the cell). The unit consists of a microprocessor system and modem for signaling to the MTX via the data line or via a traffic channel and its control unit. The supervisory unit orders the signal-strength receiver to measure the signal strength for the desired channel. The measurement result, which has a resolution of 64 levels, is transmitted to the MTX.

3.6.6 RF Test Loop

The RFTL checks the functionality of the radio equipment. On orders from the MTX, a signal path is opened over the transmitter, RFTL, antenna multicoupler, receiver and control unit, and back to the MTX. At the MTX location, it is then possible to measure the quality of the returning signal. Faults in any of the above-mentioned units will result in a too weak or distorted signal being obtained.

3.6.7 Multicoupler

A multicoupler is used to make it possible to connect several receivers to one receiving antenna. Its purpose is to distribute the signal from the antenna to the respective receiver. All outputs of the multicoupler have the same impedance as its input.

In order to avoid attenuation of the antenna signal, the antenna multicoupler is provided with active amplifiers and impedance matching networks. A schematic diagram for a multicoupler is shown in Figure 3.13.

3.6.8 Transmitter Combiner

Several transmitters can be connected to a common antenna. This is of great value, since there may be a lack of space on the masts and towers used to support the antenna system. The transmitter combining function is shown in Figure 3.14. Each combiner comprises:

- A circulator with low loss in the forward direction and high loss in the reverse direction;

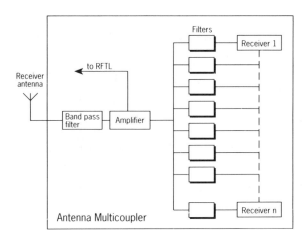

Figure 3.13 Principle for antenna multicoupler.

- High-efficiency cavity resonators filtering away other frequencies;
- Transmission line star network.

The filter-combiner provides the effect of having only one transmitter connected to the antenna at any given operating frequency. At frequencies away from the resonant frequency of the filter circuit, the transmitters are electrically isolated.

3.6.9 Autotuned Combiners

Some manufacturers offer combiners that are self-tuning and self-temperature compensating. The combiner includes a stepping motor for each channel unit that automatically adjusts the tuning. Control of the combiner filters is provided by the logic of the RFTL unit. The directional coupler feeds a broadband RF signal to the RFTL, where the signal is measured. The RFTL then sends a control signal to the self-tuning controller which controls the stepping motors.

3.7 MOBILE STATION

3.7.1 General

The mobile station can be a handheld, transportable, or vehicle-mounted telephone unit compatible with the standard cellular system air interface specification used. Subscriber equipment can also be nonmobile, serving, for instance, rural communities or oil rigs. A subscriber uses the mobile telephone almost as a fixed one, except that the mobile subscriber enters the number first, and then, after verifying the number on the display,

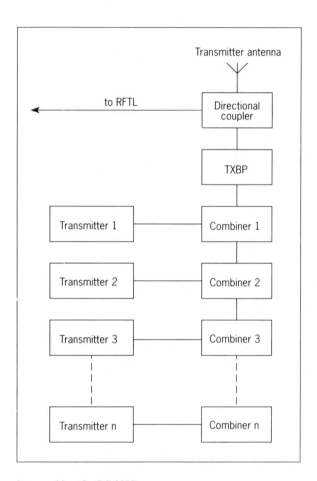

Figure 3.14 A transmitter combiner for RS 9000.

sends the number on to the base station and MTX. The mobile station, shown in Figure 3.15, consists of the following major parts.

- Transceiver (transmitter/receiver) unit for the radio communication;
- Logic unit, which handles the signaling to and from the MTX, including the setting up and clearing of calls, control of the transmitter and receiver of the radio part, and communication with the control unit;
- Handset (control) unit, including keypad and alphanumeric display; the sensing of the keys and the operation of the display handled by a microprocessor in the control unit;
- Circuits for hands-free operation (portable only);
- Duplex filter used to separate the outgoing and incoming radio channels; and
- Antenna.

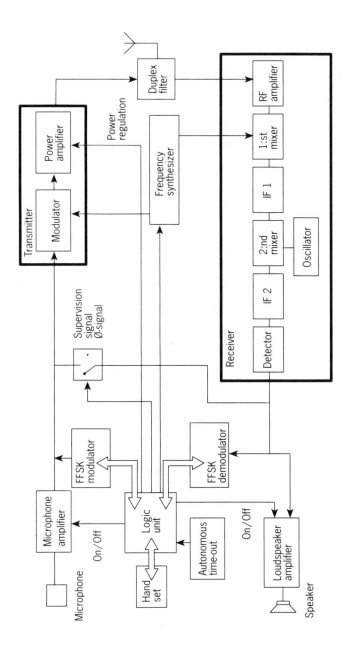

Figure 3.15 Block diagram of the mobile station.

3.7.2 Transceiver Unit

The transceiver unit consists of a receiver module, a transmitter module, a frequency generator module, and a power amplifier.

In NMT 450, the mobile station uses the frequency band 453.0 to 457.5 MHz for the transmitter, and the frequency band 463.0 to 467.5 MHz for the receiver. The distance between adjacent channels is 25 kHz and the number of duplex channels is thus 180.

In NMT 900, the mobile station uses the frequency band 890.0 to 915.0 MHz for the transmitter, and the frequency band 935.0 to 960.0 MHz for the receiver. The distance between adjacent channels is 25 kHz (with interleaved channels 12.5 kHz), and the number of duplex channels is 1,000 (with interleaved channels 1,999).

3.7.3 Logic Unit

The logic unit consist of a modem and a microcomputer system. The modem converts the digital microprocessor signals into FFSK signals. The modem is used in connection with the transmission and reception of signaling frames, but not for carrying speech information.

The microprocessor performs its tasks under the control of the programs stored in its ROM. The contents of the RAM, constituting the data of the mobile station, can be altered at any time. Examples of the contents are abbreviated numbers and information about the traffic area in which the subscriber is driving. The chief tasks of the microprocessor are control of the mobile station and signal transmission and reception to and from the MTX.

The function for control of the radio component includes switching the transmitter's output power from high to low level (and vice versa), adjusting the receiver's sensitivity, and closing or breaking the loop for supervision signals. The communication with the control unit comprises information about subscriber activities.

3.7.4 Handset Unit

The handset is the interface between the subscriber and the mobile station. The main external features of the control unit are the push-button keyboard, the alphanumeric display, a number of indicator LEDs, and symbols. The handset unit also contains a microprocessor that handles the sensing of keys, the operation of the display screen, and communications with the logic unit of the mobile station.

3.8 MARKET GROWTH

3.8.1 Nordic Countries

3.8.1.1 Sweden

Sweden is the only country in Europe where the provision of telecommunications services has always been open to competition. However, Swedish Telecom (Televerket) enjoys a dominant position in the provision of services and thus has a de facto monopoly.

The Swedish analog cellular telephony subscriber base reached 677,000 at the end of 1992. Figure 3.16 shows how the two services, NMT 450 and NMT 900, have grown in Sweden since the start of service in 1981. A similar growth pattern was experienced in each of the Nordic countries.

Comvik's rival analog cellular system has grown to about 21,000. The system is now at capacity and sales have stopped due to lack of available spectrum to expand the network further. It is generally considered that the prime reason that Comvik remains in the analog cellular market in Sweden, despite its very small market share, is to allow it access to the more financially rewarding GSM market.

3.8.1.2 Denmark

The number of Danish subscribers to the NMT 450 system, launched in 1982, currently stands at approximately 52,000. No new subscribers are being taken, since it is near capacity and existing subscribers are gradually transferring to the NMT 900 network with its greater functionality. NMT 900 is advertised as having nationwide coverage, although in practice the fringe coverage may still be patchy for hand-portable units. At the end of 1992 there were over 182,000 subscribers using the Danish NMT 900 system.

3.8.1.3 Finland

Finland remains the European country with the largest number of subscribers to a manual radio telephony system, having about 28,000 at the end of October 1991. This network is now used mainly by the residential community.

The number of subscribers to the NMT 450 system reached 180,000 by the end of 1992. (Unlike some of the other Nordic countries, subscriber growth for the NMT 450 network has continued through 1991.) Moreover, the number of subscribers to the NMT 900 system, introduced commercially in 1987, reached 196,000 at the end of 1992. These growth patterns indicate the desire of new users to have advanced features such as hand-portable phones in spite of the lower population coverage of the NMT 900 system.

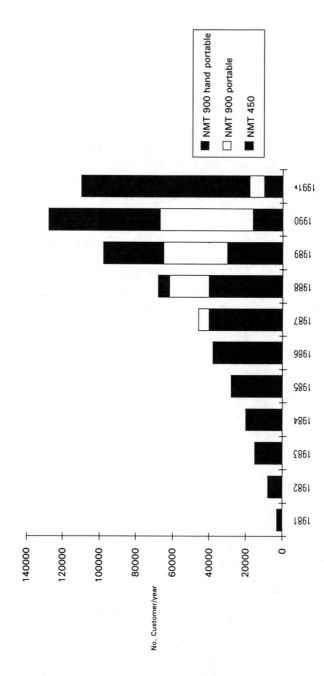

Figure 3.16 NMT 450 and NMT 900 growth in Sweden.

3.8.1.4 Norway

In common with most European countries, Norway is reshaping the provision of its telecommunication services and moving towards greater liberalization, as has been demonstrated by the recent licensing of a second GSM operator. The telecommunications operator, Norwegian Telecom, offers cellular radio services, radiopaging, and Mobitex data services.

The number of subscribers to the NMT 450 system was 154,000 at the end of 1992. The number of subscribers to the NMT 900 system, introduced commercially at the beginning of 1987, reached 133,000 at the end of 1992. NMT 450 offers nationwide coverage, while NMT 900 covers the southern part of Norway and towns, major villages, and airports in northern Norway.

The Norwegian Telecom has been lowering subscription rates to its NMT 450 and NMT 900 services to try to boost subscriber numbers in an attempt to secure the greatest return possible on its investment in analog cellular networks before they are superseded by the competitive digital ones.

3.8.2 NMT Worldwide

NMT has been adopted by many countries worldwide, including:

Andorra	Austria	Belgium
Byelorussia	China	Czechoslovakia
Croatia	Cyprus	Denmark
Estonia	Faroes	Finland
France	Hungary	Iceland
Indonesia	Latvia	Lithuania
Luxembourg	Malaysia	Morocco
Netherlands	Norway	Oman
Poland	Romania	Russia
Saudi Arabia	Slovenia	Spain
Sweden	Switzerland	Thailand
Tunisia	Turkey	Uzbekistan

Chapter 4
TACS: The U.K. Approach

E. W. Beddoes
Vodafone, Ltd.

J. R. Easteal
Alcatel Network Systems, Ltd.

4.1 BACKGROUND

Before the U.K. cellular networks started operating in 1985, the U.K. PTT (now British Telecom) had operated a mobile radio phone service for many years. This service had progressed from operator connection to direct dialing, but suffered from a severe intrinsic limitation in capacity. This was controlled only by a combination of high prices and a restriction on the total number of subscribers. Demand was such that a market in second-hand phone numbers could command a high premium.

In 1983, the U.K. Government opened the way for new mobile networks by inviting potential operators to apply for 25-year licenses. A competitive market was the intention, with the existing fixed network operators British Telecom and Mercury being restricted from direct involvement. The license applicants were required to submit business and technical plans, including proposals for the air interface specification to be used.

Of the six applicants, two operators were subsequently issued licenses. Vodafone, Ltd., a subsidiary of Racal Millicom, Ltd., owned jointly by Racal (80%), Millicom of the United States (15%), and Hambros Bank (5%). At the end of 1986, Racal Millicom became a 100% subsidiary of Racal, with Millicom acquiring equity in Racal. In 1988, Racal floated Vodafone on the U.K. stock market as The Vodafone Group plc, a completely independent public limited company. Cellnet, a 51:49 partnership between British Telecom and Securicor. Thus, British Telecom acquired a close but indirect involvement in cellular radio network operation. During 1991, the ownership changed to 60% British Telecom and 40% Securicor.

A number of conditions in the licenses were aimed directly at encouraging and ensuring vigorous competition between these new public telephone operators. Among these conditions were:

- A common air interface was required. No guidance on the choice of an air interface had been given to the various applicants, who had inevitably produced differing proposals. A common air interface would allow subscribers to transfer from one competing network to the other without the penalty of replacing their mobile terminal if they were dissatisfied with the service that they received from their network operator. The network operators had to jointly propose the specification for Government approval.
- The licensees were to provide service to 90% of the U.K. population by 1990. This ensured that they could not simply offer service in high-density areas, but must provide substantially national coverage. In practice, each achieved this coverage by mid-1987. By 1991, both networks provided cellular coverage to greater than 95% of the population.
- Network service to individual and corporate users could only be offered to subscribers indirectly through service providers. This would ensure even more competition at the retail level. Service providers, and any retail organizations that they dealt with, could offer differing packages of equipment, connection, and call charges.

Against this background, Cellnet and Vodafone eventually agreed on a modified version of the AMPS specification that was in use in the United States (see Chapter 2). This revised AMPS specification was published as the TACS [1] specification after copyright clearance had been agreed on with the EIA.

Subsequently, further modifications have been made to increase the spectrum allocation (and hence the number of channels) and enhance the signaling and overcome fraudulent use. These have resulted in ETACS [1] and TACS 2 [2], respectively. The ETACS frequency extension is discussed in Section 4.2.3, and the TACS 2 modifications are described in 4.2.7.

4.2 TACS—OUTLINE SPECIFICATION

4.2.1 Requirements

The specification adopted in the United Kingdom had to attempt to satisfy a number of requirements, some mandatory, others strong preferences. These requirements were a mixture of administrative, technical, and commercial, namely:

- The European mobile radio frequency allocation of 862 to 960 MHz;
- A channel spacing based on 25-kHz multiples, or submultiples, to follow standard European practice;

- Rapid start of the service, to satisfy a heavy demand from business;
- Low technical risk, implying application of an existing standard;
- Strongly supported by equipment suppliers.

4.2.2 Air Interface

The TACS air interface was largely produced (and is currently maintained) by the U.K. network operators working within a technical group, now named TG (TACS), of the U.K. Department of Trade and Industry (DTI). As described previously, the core of the specification was the U.S. AMPS specification as it existed in 1983. This had a number of well-recognized limitations which were overcome in the TACS version. Inevitably, in what has developed into extremely complex networks, other problems have been identified, resulting in the TACS 2 revision.

4.2.3 Frequency Allocation

Originally, the bands 890 to 915 MHz (mobile transmit) and 935 to 960 MHz (mobile receive) were defined with the intention of allocating blocks of channels within this band as circumstances demanded. In practice, however, only the lower 15 MHz were released for TACS use in the United Kingdom. The remaining 10 MHz were held back by the DTI for use by the same operators in anticipation of the pan-European GSM system. In addition, a lower adjacent band has been defined for ETACS: 872 to 890 MHz (mobile transmit) and 915 to 935 MHz (mobile receive). Of this lower band, the top 2 MHz have not been released; they are reserved for use by other services, such as the digital short-range radio (DSRR) concept. Figure 4.1 shows the TACS and ETACS frequency allocations.

4.2.4 Transmitter Power

To help minimize the interference potential of cochannel mobiles, their transmitter power can be dynamically controlled by the network. This control is based on signal strength

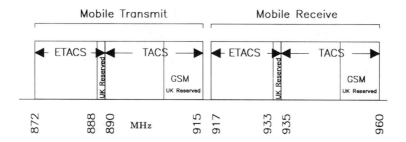

Figure 4.1 TACS and ETACS frequency allocations.

and signal quality as determined by the base station. To allow the network to make controlled use of this facility, the power levels are defined in 4-dB steps. The maximum power of specific equipment is designed by the manufacturer from one of four permitted classes. The maximum power to be used by any mobile in a specific cell is broadcast on the base station control channel, while the actual power of the mobile during a call is dynamically controlled by the base station in use. Figure 4.2 shows the power classes and power control steps.

To further minimize interference levels and to conserve portable battery power, discontinuous transmission is permitted; however, this was only implemented by Cellnet in the United Kingdom

4.2.5 Modulation

TACS uses FM for speech and FSK for signaling. The voice channel deviation of ± 9.5 kHz is wider than the ± 5-kHz used by PMR for 25-kHz channel spacing and leads to increased adjacent channel interference. This typifies the system design differences between range-limited PMR systems and interference-limited cellular systems. The extra deviation, however, provides better call quality with improved cochannel interference performance, and by careful allocation of channels between cells, adjacent channel interference is minimized, as shown in Figure 4.3.

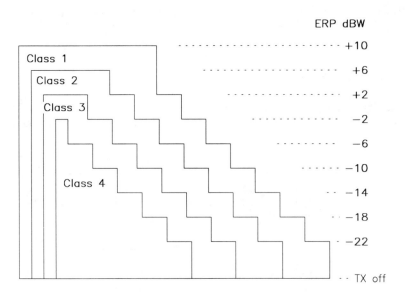

Figure 4.2 TACS Mobile transmit power classes and steps.

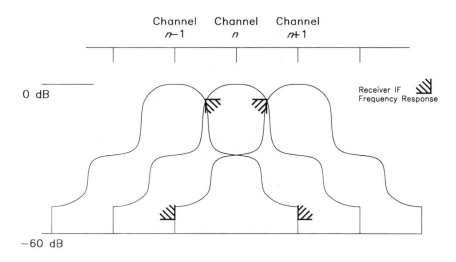

Figure 4.3 TACS channel modulation mask showing adjacent channel interference.

4.2.6 Receiver Sensitivity

To ensure a minimum performance in the network, the receiver sensitivity is specified at −113 dBm for 20-dB SINAD. Additionally, despite being an FM receiver, the mobile must measure the received signal strength of control channels to close tolerances to allow correct selection of the local serving cell. This necessitates a fairly stringent measurement linearity specification; in practice, a maximum nonlinearity of 2 dB is allowed.

4.2.7 Signaling

Four signaling paths are used in TACS as in AMPS, two forward channels towards the mobile and two reverse channels from the mobile. Call setup is carried out on separate control channels: the forward control channel (FOCC) and reverse control channel (RECC). Once the call has been allocated a voice radio channel, signaling is carried out on the forward voice channel (FVC) and the reverse voice channel (RVC).

Signaling is carried at 8 Kbps as needed on the FVC and RVC, speech being blanked during bursts of data. In practice, the burst is audible at low levels but rarely affects the intelligibility of the call. Error coding and majority voting of repeated data are used to maintain high data integrity on all signaling paths.

In addition to the 8-Kbps data paths, the base station transmits and the mobile repeats a SAT on the voice channel. The particular tone used by a specific base station is signaled to the mobile on a control channel as the digital color code. The SAT is one of three frequencies (5,970, 6,000, and 6,020 Hz) and is used to provide additional identification of mobiles to distinguish them from others using the same radio channel elsewhere in the cellular reuse pattern. When an interfering mobile is dominating a base station receiver, a "crossed-line" effect results. The base station can recognize the mobile with the incorrect SAT and the call should be terminated. Table 4.1 gives an activity summary of the signaling protocols used in TACS [3].

4.2.8 TACS 2

There are a number of operational problems with the TACS system, some of which are attributable to weaknesses such as insufficient digital color codes, number of SATs, lack of identification of a mobile at handoff and release, and the relationship between control channel selection rescans and registrations in the basic TACS (and AMPS) specification. For this reason, the TACS specification has recently been modified (to become TACS 2) to overcome some of them, at least for new mobiles. This helps to reduce the cochannel signaling problems, which otherwise result in false releases and false handovers, a significant cause of dropped calls.

The changes introduced in TACS 2 all affect the signaling specifications. They have been defined with care to accommodate a mixture of TACS 1 and TACS 2 mobiles and base stations within the networks and to allow operation at the lowest common denominator protocol for any individual call. These changes have overcome a number of protocol deficiencies which, in the past, produced dropped-call occurrences. They have also reduced the signaling load on the network and have replaced the 8-kHz tone on/off signaling with 8-Kbps data for on/off-hook status indication.

At the same time, automatic call repeats have been limited, the DTMF signaling has been redefined to improve its reliability, and the number of dialed digits permitted increased from 16 to 32.

The susceptibility of the network to fraud from cloned mobiles has also been addressed. TACS 2 introduces an authentication facility which uses a *pin* number with the telephone number and ESN as parameters in a confidential algorithm to protect subscribers and operators alike.

4.2.9 Switching Network

The telephony switching environment for cellular radio systems has evolved in two directions. First, the radio controller used in the PMR system has been enhanced to produce small switches of 100 to 200-erlang traffic capacity. The second alternative has emerged from the established fixed network telephony switch manufacturers, who have migrated their trunk switch offerings to support the mobile telephony role, generally yielding a higher traffic capacity of 1,000 to 2,000 erlangs. Given the number of subscribers in the United Kingdom and the future potential market, and since the busy-hour traffic level per subscriber has historically been around 20m erlangs, it should come as no surprise to

Table 4.1
Signaling Protcols Used in TACS

System Activity	Signaling Channel	Mobile Activity
Power-up		
		Mobile reads the fixed data in its memory. (System/control channel positions)
System overhead parameters and other overhead messages	FOCC→	Scan FOCCs. Read FOCC information (channels available/DCC/power level/facilities available) and store. Remain in monitor mode of FOCC
Call Initiation		
System overhead parameters and messages	FOCC→	Monitor mode of FOCC
		Mobile user loads required number into the mobile and activates the Send function. Mobile checks its system access data, monitors the busy-idle stream, and performs a system access
In response to the origination message, the system changes the busy-idle bits to busy within approximately 1 to 4 ms	←RECC	Origination message sent (contains mobile ID, serial number, called number)
On receipt of the complete origination information, the switch processes the called number, checks the mobile status–valid on system and confirms the base station used		Monitor FOCC for mobile control message
Generate channel assignment (mobile control message). Dial out required number. Connect selected outgoing trunk to the assigned voice channel.	FOCC→	Receive voice channel assignment, power level, SAT and store in memory
On receipt of SAT, connect audio path	←RVC	Check memory to confirm if channel is in assigned group. Retune to voice channel. Send SAT
	←Conversation→	Mobile user now hears call progress tones associated with called number and awaits answer

Table 4.1 (continued)

System Activity	Signaling Channel	Mobile Activity
Call Reception		
System overhead parameters and other overhead messages	FOCC→	Monitor FOCC for mobile control messages
Receive incoming call. Generate page (mobile control message)	FOCC→	Receive page. Monitor the busy-idle stream and perform a system access
In response to the page response message, the system changes the busy-idle bits to busy within approximately 1 to 4 ms. Verify the mobile data received and confirm the base station used	←RECC	Send data (mobile ID no.) in response to page
Generate channel assignment (mobile control message)	FOCC→	Receive voice channel assignment, power level, SAT and store in memory
		Check memory to confirm if channel is in assigned group
Receive confirmation. Associate land party with assigned channel. Send alert order to mobile	←RVC	Retune to voice channel. Send SAT
	FVC→	Alert device activated
Received ST confirms alert	←RVC	Send ST
		Handset lifted to answer, alert terminated, ST off
Removal of ST detected and audio path connected	←Conversation→	Conversation commences
Power Level Change		
	←Conversation→	
Monitor signal level. If level too low, send increase power level order to mobile	FVC→ (mute audio)	Receive change power level order
Receive order confirmation. Continue monitoring signal level.	←RVC (mute audio)	Store new level. Send order confirmation
Handover		
Monitor signal level. If the level is too low and the mobile is on the max. power level, check adjacent base station received power levels and identify a free channel at an appropriate base station	←Conversation→	

Generate new channel assignment for handoff. Bridge audio path via switch to new base station.	FVC→ (mute audio) / No channel in use	Receive handover order, store data (new channel no./SAT/power level). Send ST for 50 ms. Turn off transmitter
On receipt of SAT, remove audio path to old base station. Return to monitor signal level	←Conversation→	Retune, set power level, new SAT. Turn on transmitter

Cleardown

System detects long duration ST indicating cleardown and releases the connection to the other party	←Conversation→ / ←RVC / No channel in use	The user decides to terminate the call and replaces the handset. ST is sent for 1.8 sec
System parameter overhead message generation	FOCC→	Mobile retunes to the control channels and scans for strongest channel. Checks data against those in store and modifies if necessary

Additional Service Request

Monitor signal level	←Conversation→	The user loads the type of additional service required, and the additional number if applicable and activates Send
Receipt of ST indicates a Flash Request. Generate a Send Address Order	←RVC	ST is sent for 400 ms
	FVC→ (mute audio)	Receipt of Send Address order. Send Called Address message
Process the received data and provide facility as applicable	←RVC (mute audio) / ←Conversation facility feature→	

Note: Courtesy of the U.K. Department of Trade and Industry.
*Used to provide access to three-party calls, call diversion, and so on.

learn that the U.K. operators use the second alternative of larger switches, each of which supports between 50,000 and 100,000 subscribers.

The need to provide national coverage initially resulted in a number of switches being strategically placed across the country at centers of traffic in order to optimize the cost of cell-to-switch transmission links purchased from British Telecom or Mercury. As traffic has increased, additional switch sites have been acquired to accommodate the MSC, of which 23 were in use in the Vodafone network in 1990, and the number is still growing. These switches have been distributed across 14 sites, as shown in Figure 4.4.

Initially, switches were fully interconnected to all of the others in the network; however, in order to create a manageable network and minimize link costs as the number of switches has increased, a two-tier approach has been adopted by both operators by the creation of an overlay TSC network. MSCs are connected to at least two TSCs for security,

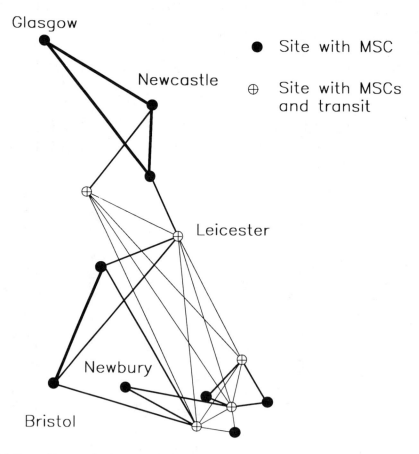

Figure 4.4 Geographic network of Vodafone switching sites.

the TSCs being fully meshed. The schematic switching network for Vodafone is shown in Figure 4.5.

4.2.10 Mobility

A mobile switching network is often regarded as one of the first large-scale implementations of an intelligent network. This is because it has to manage subscribers' mobility by routing calls to them correctly as their location changes, and by dealing intelligently with mobiles that are out of coverage.

Subscriber data were stored on a database that forms part of each MSC. Initially, attempts were made to allocate numbers to subscribers based on judgment as to where they would make the majority of their calls (e.g., their home area or workplace). For this purpose, *subscriber number allocation zones* (SNAZ) were defined, the objective being to minimize the amount of signaling necessary for roaming in the network between MSCs. It soon became apparent that the management of such a system would not be possible, since the service providers (retailers of equipment and air time), who had been allocated blocks of numbers, would distribute them to their nationwide dealers. In addition, the volume of subscribers using the system in, for instance, London would exceed the capacity of the London switch database, whereas the database in Glasgow, for example, would have plenty of capacity.

Universal roaming within the network was thus accepted and the network design philosophy adjusted accordingly. In a subsequent development, subscribers have been distributed evenly across databases implemented on VAX computers, and the network has been dimensioned in recognition of the fact that a high percentage of subscribers are roaming most of the time.

Thus, when a U.K. TACS mobile phone is switched on, its data will be retrieved from an HLR somewhere in the network and stored in a VLR on the switch serving the cells in the area where the phone is currently located. The HLR will note the identity of the current VLR and the fact that the mobile is active. Incoming calls for the mobile will interrogate the HLR based on knowledge of the mobile's number and where each number range is stored. If the mobile is deemed active, the call will be routed to the appropriate VLR for paging the mobile. This is shown in Figure 4.6. Mobiles will also re-register periodically (typically every 15 min) to let the system know they are still active. If registration is not successful, they will be marked inactive by the system after a period of 5 min and thus will not be paged. This 5-min delay allows mobiles only temporarily out of coverage to re-register successfully.

Mobiles also re-register when they cross from one switch area to another, resulting in the cancellation of the first VLR entry and the creation of a second VLR entry on the switch serving the new area. In this way, calls can be correctly routed to the mobile as it moves from one location area to another.

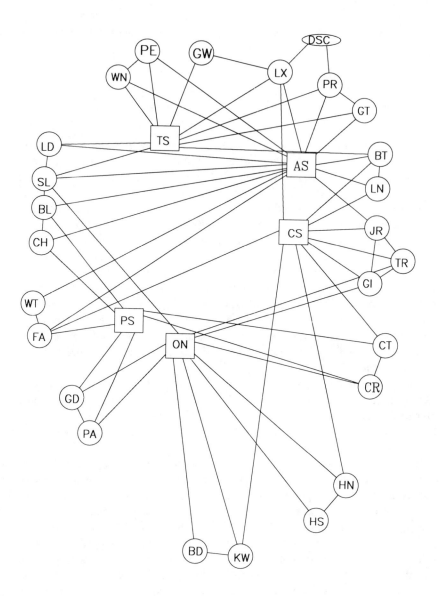

Figure 4.5 Vodafone switching network (transit and cell site links omitted).

4.2.11 Interconnection

Mobile network operators in the United Kingdom are licensed PTOs having similar rights and obligations to those of British Telecom and Mercury, but with respect to the provision

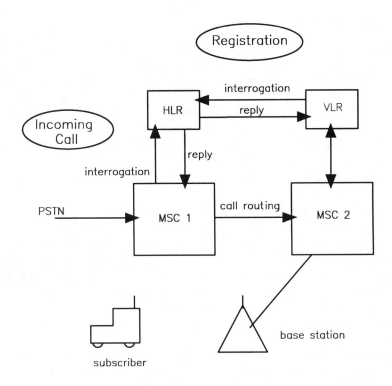

Figure 4.6 Incoming call HLR and VLR messages.

of mobile telephone services rather than fixed telephones. A requirement common to all is the right of connection to other PTO networks.

Interconnection between the mobile networks and British Telecom occurs at the MSC and TSC levels. All switches are usually connected to two digital main switching units (DMSU) in the British Telecom network for security; all routes are both-way digital using CCITT No. 7 signaling to British Telecom's specification. Mobile switches are located, if possible, in high-traffic areas to minimize the transmission costs with base stations and to enable British Telecom to dispose of the considerable mobile traffic delivered to the fixed network.

Links also exist to the Mercury network, and there are direct links between Cellnet and Vodafone, the latter carrying surprisingly high levels of traffic between these communities of interest.

The U.K. cellular networks are not insignificant in telecommunications terms. Many of the techniques specified in GSM for international roaming have been in use in the United Kingdom for many years and have provided valuable experience with the problems (and solutions) of operating large integrated networks. Much of this experience has been brought into the GSM specifications.

4.3 U.K. EXPERIENCE

The capacity of a cellular system and its ability to cope with a given traffic density are often taken to mean the same thing, since the system must be able to accommodate the peak traffic density likely to be encountered, all other lower density areas by definition being easier to deal with. The inability of a system to deal with the maximum traffic density is often referred to as the *capacity limit* even though the total system capacity could easily be increased in other areas where there is no unsatisfied demand. There are three fundamental factors that limit the ultimate capacity that can be provided; they are:

- The total amount of spectrum available;
- The cochannel interference level that can be tolerated by the radio equipment for acceptable call quality;
- The size (smallness) of the cells that can be established.

The availability of spectrum has already been discussed. It is worth noting that with analog FM systems, attempting to increase system capacity by reduction of the equipment channel spacing results in a lowering of the protection against cochannel interference, necessitating a higher integer cell cluster repeat pattern. The capacity gain achieved by a reduction in the channel spacing is therefore offset by an increase in cluster size. Thus, comparisons of the number of channels per megahertz of NMT 900 (12.5-kHz channel spacing) with TACS (25-kHz channel spacing) are unrewarding, since it is the spectrum allocation that is fundamental.

A theoretical analysis shows that TACS requires a C/I ratio of about 18 dB; although in practice this can be reduced to 12 dB and still give acceptable performance.

Cellnet has used the 6-dB margin to implement a four-site repeat pattern, which requires four sites for a given capacity at the expense of a slightly lower quality of service. Vodafone has used this C/I ratio margin to allow sites to be placed closer together and to cope with irregular clusters of cells of different sizes, thereby allowing sites to be brought on as they become available, rather than necessitating a "big-bang" approach to roll out and expansion. A practical dimension to this problem is the difficulty of finding cell sites at a precise location in urban environments where the site spacing can be as low as 1 km. Site location inaccuracy inevitably leads to a dilution of the potential increase in capacity of reduced cell size.

Each method of using the margin leads to an increase in capacity; the theoretical benefits of either system are degraded by practical difficulties and imperfect site location. In practice, it is unlikely that either solution has a real advantage over the other, and each probably has a marginal advantage in specific situations.

4.4 CELLULAR RADIO IMPROVEMENTS

4.4.1 Overlay Cells

As subscriber demand increased, it became necessary to explore additional methods of increasing capacity. Site spacings in London were reduced to 1 km to cope with peak traffic densities of around 32 erlangs/km^2. (It is worth noting that the resulting cell sizes are as small as those being proposed for the personal communications network (PCN) service.) A typical cell plan is shown in Figure 4.7. Further research effort was therefore directed towards methods of increasing the density of traffic that could be handled. This led to an evaluation of the use of overlaid cells and microcells.

The overlaid cells concept is shown in Figure 4.8. It allows further reuse of frequencies at each site, provided they are only used by mobiles within a smaller radius than that of the macrocell, thus providing additional capacity at the center of a cell. It is important that the cell site is located where the peak traffic occurs. The overlaid cells use frequency groups that are already allocated to normal adjacent macrocells. The reuse distance for the overlaid cell appropriate to a seven-cell cluster can thus be maintained, and hence the quality of service is approximately the same as that of the main cell plan.

Inevitably, there is a price to pay for such improvements; in this case, the system complexity is increased. Software is required in the base station to recognize mobiles that access the site at a high signal level; these mobiles are then allocated channels in the overlaid cell in preference to the macrocell on the assumption that they are close to the base station. The macrocell channels are reserved for mobiles farther away or for when all overlaid cell channels are fully occupied; intracell handover is, however, now required. Using this technique, subscriber densities of around 60 erlangs/km^2 can be achieved without noticeable reduction in service quality.

4.4.2 Microcells

The alternative of using microcells has also been used. A microcell may be defined, without using the concept of cell size, as a cell in which the antenna is mounted below the average building height, amidst the "clutter." Because of the additional clutter and consequent loss of signals from the macrocell base station in this environment, the microcell can reuse a frequency group without degradation of service quality. Microcells have thus far only been used to cover traffic "hot spots" in urban areas rather than as a homogeneous solution. The microcell technique allows traffic densities of up to 100 erlangs/km^2 to be supported.

4.4.3 Other Techniques

A number of other techniques have been developed to cope with the nonuniformity of practical cellular systems and to maintain and improve the quality of service (and hence capacity) in a real environment.

A major potential problem can be caused by dissimilar antenna heights, which range from 60m, the height chosen for large cells at start of service, to 15m, the average height used in a mature small-cell system. In these cases, the high site can capture all the traffic and run into congestion, while the surrounding, lower sites have spare capacity. This can

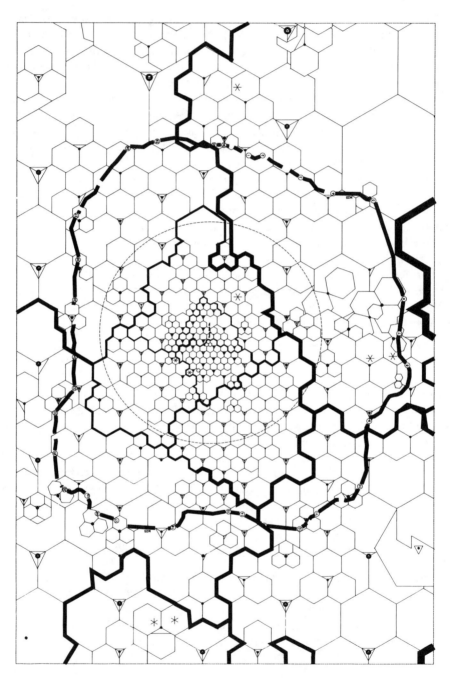

Figure 4.7 London cell plan.

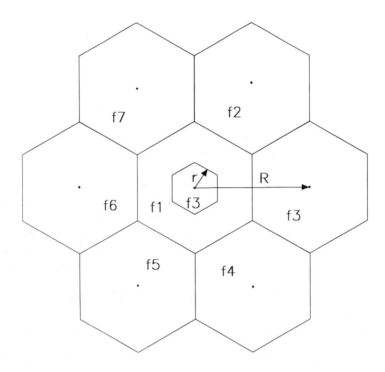

Figure 4.8 Overlaid cell concept. Overlaid cell reuse ratio = *R/r*.

be remedied by redirecting traffic from the congested high site to the surrounding sites when few channels are available, thus redistributing the traffic and providing a perceived grade of service determined by the capacity of the surrounding sites.

In some circumstances, this situation may be further remedied by tilting the antenna downwards by up to 10 deg to make use of the cutoff of the vertical beam of the directional antennas used, in order to restrict coverage and interference. Typical 3-dB antenna vertical beamwidths are 5 deg.

Another technique used to reduce interference levels and to ensure that the mobile is on the optimum cell and operating at the lowest power is now in common use and often referred to as *midpoint handoff*. The base station and switch parameters have been modified so that when the signal strength of a mobile falls below a predetermined threshold, instead of requesting a mobile to increase its transmit power, the surrounding sites are first examined to determine if there is a better cell; if not, the mobile power is turned up instead. The overall level of interference is thus reduced because the mobiles run at a lower average power level and because they are in the correct cell coverage area for more of the time.

Cellular planning techniques have thus moved on from the early implementations used in 1985. These developments have been essential to meet subscriber expectations of coverage, accessibility, and service quality. Chapter 11 describes how planning tools can be used to maximize system performance.

4.4.4 Switching Improvements

In Section 4.2.9, the use of HLRs and VLRs was discussed as a mechanism for national roaming and a balanced distribution of subscriber data. In a large high-density network with many switches, life is not that simple. The mobiles in the boundary areas between switches not only frequently change their location areas, but also may make dual accesses, causing registration in two location areas.

It was essential to find a solution to this problem, since, in the current networks, the number of switches (and hence boundaries) is very much higher than originally envisaged. For example, the number of switches serving the area inside the M25 (the London ring road) today is approximately equal to the number of cell sites serving that area when the system first opened in 1985! The number of boundaries can be seen in Figure 4.7. Therefore, the mobile may register with and thus be paged on the wrong switch. Considerable effort has been expended to identify all the possibilities and resolve this problem. Much software has been written to provide a solution, and some of the weaknesses have been resolved by changes in TACS 2. One solution, for example, is multiswitch paging, but this has the disadvantage of using more paging channel capacity and requiring longer call setup times.

Full mobility management and personal numbering thus exists within the U.K. analog mobile networks. Mobility across such networks and, to a greater degree, within GSM networks and between the mobile and fixed networks (with personal numbering to allow a subscribers to change their network) can be based on the same principles of distributed databases and smart cards and should be the way forward for personal numbering.

Networks for mobile phones, however, have other unique problems when compared to fixed networks. A mobile may be unavailable for much of the time, either because it is out of the coverage area or because it is switched off. Mobile networks are thus more intelligent than their fixed counterparts in this respect, calls defaulting to a recorded announcement if the call cannot be completed. A more useful response is obtained if the subscriber uses a call diversion service on "busy" or "no reply;" such diversions are often set to route the call to the home or the office or to a centralized personalized voice messaging (mailbox) facility provided by the mobile network operator. This mailbox facility can dial out to a pager to alert the user that a message has been left, or to call back the mobile automatically when it becomes active, thus ensuring that contact can be made.

4.5 COVERAGE

The initial requirement for "rollout" of a cellular system is radio coverage. This is even truer today for new operators than it was when the current cellular networks were opened in January 1985. There is a critical mass in terms of radio coverage at which subscribers realize they can call from most places they find themselves in. At this stage, the network becomes most useful to subscribers and they begin to depend on it. This experience has been repeated with the introduction of the NMT 900 networks in Scandinavia.

In the United Kingdom, the operators had an obligation under the terms of their licenses to provide coverage for 90% of the population by 1990; this was achieved much earlier, by mid-1987.

Quality coverage is, however, required in order to be able to offer reliable service, and this can only be achieved by careful site selection, design, and verification. Fortunately, it is now the norm to use propagation and terrain modeling techniques to calculate the radio coverage from a given site [4]. This technique has been employed by U.K. cellular radio operators in the design and rollout of the current networks, using initially a ground height terrain map of the United Kingdom with 500m horizontal and 1m vertical resolution. Propagation models based on those of Okumura [5] and Edwards and Durkin [6] have been used with some success, the measured coverage for suburban and urban cells being close to that predicted. But these have more recently been overtaken by the development of more sophisticated models capable of working with resolutions down to 50m. An allowance is made in the calculations for clutter and for log normal shadowing, so that the probability of coverage within the defined area is greater than 90%.

For urban areas and small cells, accurate prediction of coverage is more difficult, and considerable effort has recently been put into improving the modeling techniques, the accuracy of the clutter data, and the resolution accuracy of the terrain data. These developments have been driven by the small-cell structures now required in urban areas in TACS and by plans to introduce a microcellular communications network (MCN), based on GSM technology, for the consumer/domestic market,.

Current systems, however, require that a comprehensive coverage map of the United Kingdom defining primary and secondary coverage should be available for use by the subscriber. This is achieved by entering the coverage data from all in-service sites onto a database, which is then used as the source of data for a printed coverage map.

An interesting low-cost development has combined the coverage data with the well-known "Autoroute" package for PCs, thereby allowing users to assess the coverage in their town or village or on any planned journey. This package is now used by engineers and service providers alike as a convenient method of providing up-to-date coverage information for subscribers.

Radio coverage planning has thus made tremendous advances in the past 5 years owing to the need to define cellular coverage accurately and precisely by the use of propagation modeling and terrain databases. The additional spur has been the sheer impossibility of economically assessing site options in high volume by recourse to radio surveys.

4.6 U.K. OPERATING CHARGES

Cellnet and Vodafone have maintained their charges at the same price since they opened in January 1985. Thus, with inflation, the real costs have dropped significantly. Significant price competition has not yet occurred; this may be about to change with the appearance of two PCN operators in the market.

Both U.K. operators have now introduced a second tariff structure, which is aimed at the sort of subscribers who make few calls but are willing to pay a premium when they do.

An interesting feature to note is the decline in average revenue per subscriber since the start of cellular. This probably reflects a change in user profile from senior staff to lower staff levels who operate with tighter budget constraints. The charges in early 1993 for each network are listed in Table 4.2.

Deciding to which operator and to which of its tariffs to subscribe demands careful study.

Table 4.2
Cellular Rates Charged by Cellnet and Vodafone

	Cellnet London	*Cellnet Provincial*	*Vodafone London*	*Vodafone Provincial*
Premium				
Connection	£65.00	£65.00	£50.00	£50.00
Monthly	£25.00	£25.00	£25.00	£25.00
Peak per minute	35p*	25p*	33p†	25p†
Off-peak per minute	12p	12p	10p	10p
Low use				
Connection	£20.00	£20.00	£30.00	£30.00
Monthly	£14.89	£14.89	£15.00	£15.00
Peak per minute	50p‡	50p‡	54p◊	46p◊
Half-peak per minute			27p§	23p§
Off-peak per minute	20p	20p	15p	15p

*Monday–Saturday 0800–2200
†Monday–Saturday 0730–2130
‡Monday–Friday 0800–1900
◊Monday–Friday 0700–2000
§Saturday–Sunday 0700–2000

4.7 TACS NETWORKS

TACS and ETACS networks are operational today in a number of countries worldwide. The major systems are detailed in Table 4.3.

4.8 CONCLUSION

The United Kingdom has seen an unprecedented growth in demand for public mobile telephony since the service began in 1985. This demand has driven the network operators

Table 4.3
TACS and ETACS Networks Worldwide

Country	System	Subscribers
U.K., Cellnet	ETACS	634,970
U.K., Vodafone	ETACS	820,940
Ireland	TACS	42,900
Italy	TACS	768,100
Spain	TACS	133,860
Austria	TACS	120,222
China	TACS	160,000
Macao	TACS	10,000
Kuwait	TACS	42,500
Pakistan	TACS	16,000
United Arab Emirates	TACS	36,000

Note: Other countries employing TACS systems include Malaysia,
Singapore, Bahrain, Mauritius, Sri Lanka, and Malta.
Source: *Mobile Communications*, January/March 1993.

to develop original technology in both radio and switching techniques, breaking new ground in terms of network capacity and subscriber mobility. At present, networks operate with cell sizes as small as 1 km and with intelligence greater than that of the fixed network. This has resulted in networks that offer a better quality and range of intelligent services and that handle a larger number of subscribers than any other mobile network in the world today. Many of the lessons learned have been included in the ETSI-GSM recommendations for the pan-European second-generation digital cellular standard.

REFERENCES

[1] Chapman, I., and E. Frazer, "A Guide to TACS-2," *International TACS Conference,* 1989.
[2] *Total Access Communication System, Mobile Station–Land Station,* Compatibility Specification, U.K. Department of Trade and Industry.
[3] *A Guide to the Total Access Communications System,* U.K. Department of Trade and Industry, 1985.
[4] Hall, M. P. M., and L. W. Barclay, eds., *Radiowave Propagation,* London: Peter Peregrinus, 1989.
[5] Okumura, Y., E. Ohmorj, T. Kawano, and K. Fukuda, "Field Strength and Its Variability in VHF and UHF Land Mobile Service," *Rev. Elect. Comm. Lab. (Japan),* Vol. 16, 1968, pp. 825–873.
[6] Edwards, R., and J. Durkin, "Computer Prediction of Service Area for VHF Mobile Radio Networks," *Proc. IEE,* Vol. 116, No. 9, 1969, pp. 1493–1500.

Chapter 5
Analog Cellular Radio in Japan

M. Sakamoto

NTT Mobile Communications Network, Inc.

Japan introduced its first cellular system in 1979. Since that date growth has been rapid and new systems have been introduced to cope with the demand. This chapter describes the development of the analog cellular systems in Japan, describing its technical, operational, and commercial aspects.

5.1 BACKGROUND

Cellular radio was introduced in Japan on 3 December 1979. Before then, there were no public land mobile telephone systems, and neither was there a strong demand for a public mobile service. This was partly because there were many pay telephones in this small, overpopulated country. Radio paging, however, was introduced early in 1968, and in 1979 the subscription level reached more than 900,000. The service worked well with the aid of the densely distributed pay telephones. Another major mobile system before cellular was a maritime mobile telephone system, which was introduced in 1964 and extends approximately 50 km from the coasts of Japan. The number of subscribers exceeded 10,000 in 1979.

The Electrical Communications Laboratories of NTT had started research into land mobile telephone systems in 1953, and its development was completed in 1967. The system was noncellular but fully automatic, and its frequency band was 450 MHz. This mobile telephone system, however, was not put into commercial service because of a predicted lack of available frequency bandwidth in the 450-MHz band [1].

Paralleling the development of this system was the intensive study of mobile propagation in the 450- to 2,000-MHz bands. The results, which are well known as the *Okumura curves,* have been the basis of most subsequent land mobile radio design [2,3]. Later, in

1980, the Okumura curves were formulated as propagation loss, making it possible to put the formula into calculations [4].

NTT Laboratories began research on cellular systems in the early 1960s. The first report appeared in 1967 and discussed fundamental problems of cellular systems [5]. From around 1967, efforts towards the development of an 800-MHz band cellular system were accelerated. The targets of the system were:

- High capacity (100,000 subscribers in an area);
- Nationwide, fully automatic operation (direct dialing, location registration, pursuit switching, etc.);
- Continuity of communications (handover).

In 1971, the first technical trial system was developed and field trials were carried out in the Tokyo metropolitan area. In 1974 and 1975, the second and final technical trial system was developed, and indepth testing was successfully completed.

5.2 JAPANESE SYSTEMS

5.2.1 Overview

The operators in Japan are NTT, Nippon Idou Tsushin Corp. (IDO), and the seven subsidiary companies of Daini-Denden, Inc. (DDI). They operate six cellular systems. IDO is a radio common carrier that operates cellular systems in the Tokyo and Nagoya areas. DDI is a long-line carrier and has seven subsidiary companies that operate cellular systems in areas other than those controlled by IDO. The six systems include NTT's first-generation cellular (mobile control station (MCS)-L1), NTT's second-generation cellular (MCS-L2), DDI's JTACS, DDI's and IDO's NTACS, and IDO's system design, which is based on NTT's MCS-L2. The systems and regions in which they operate are shown in Table 5.1.

The first system in Japan, NTT's MCS-L1, was introduced in 1979. Several years after its introduction, the number of subscribers began to expand. The traffic distribution

Table 5.1
Japanese Operators and Their Systems

Operator	Region	System
NTT	Nationwide	MCS-L1
		MCS-L2
IDO	Tokyo and Nagoya, including their surrounding areas	MCS-L2-based NTACS
DDI's subsidiaries	Areas other than IDO areas	JTACS
		NTACS

within a service area was found to be extremely concentrated towards the center of cities, as illustrated in Figure 5.1. The traffic concentration reduces geographical frequency reuse efficiency. With these two facts threatening to overload the system, NTT started research and development on a second-generation higher capacity cellular system. The system was designed emphasizing an increase in the number of radio channels rather than improvement in geographical reuse efficiency [6]; that is, the channel separation was narrowed to one-half that of MCS-L1.

In April 1985, Japan's telecommunication services were liberalized, and new cellular carriers, IDO and DDI's subsidiaries, vigorously competed for the cellular market that had previously been monopolized by NTT. The Telecommunications Technology Council of the Ministry of Post and Telecommunications, after discussing the merits of various systems, released a report in April 1986 stating that the systems should be based on either NTT's high-capacity system (MCS-L2), the North American system (including the U.K. system), or the Nordic system. In accordance with this report, NTT introduced MCS-L2 in May 1988, and IDO introduced their MCS-L2-based system in December of the same year. DDI's first system, JTACS, based on the U.K.'s TACS system, was put into service in July 1989 by the Kansai Cellular Telephone Company.

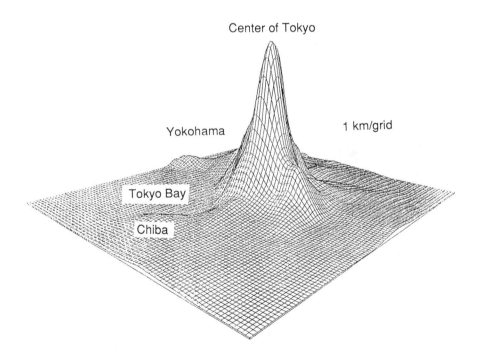

Figure 5.1 Traffic concentration.

In April and October 1991, DDI and IDO introduced Motorola's NTACS, which narrows channel separation to 12.5 kHz with interleaving, a one-half reduction over that of JTACS.

The frequency bands for the Japanese analog cellular systems are shown in Figure 5.2. All are in the 800-MHz band. NTT uses 15 MHz × 2 for both MCS-L1 and MCS-L2. The service areas of IDO and DDI do not overlap, and they use common bands of 12 MHz × 2.

5.2.2 NTT System

MCS-L1 was the first commercial cellular system in the world. The major features are summarized in Table 5.2 and the network is shown in Figure 5.3 [7].

The frequency band is a pair of 15-MHz bands: 870 to 885 MHz for base transmission and 925 to 940 MHz for mobile transmission. The mobile unit has 600 channels, the separation of which is 25 kHz. Signaling in the radio path is 300 bps biphase (Manchester).

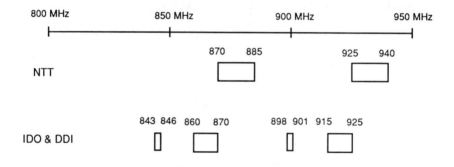

Figure 5.2 Frequency allocation for analog systems.

Table 5.2
Main Features of MCS-L1

Item	Feature
Frequency band	800 MHz
Transmit and receive separation	55 MHz
Channel separation	25 kHz
Number of channels	600
Modulation	PM (maximum deviation 5 kHz) with 2:1 compandor
Signaling	300 bps biphase
Control Channels	Paging channel and access channel with multitransmitter simulcasting

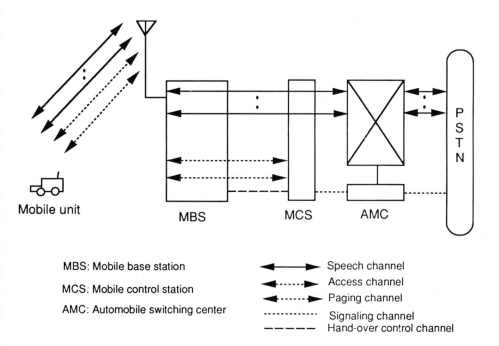

MBS: Mobile base station   ◄────► Speech channel

MCS: Mobile control station   ◄┄┄► Access channel

AMC: Automobile switching center   ◄┄┄► Paging channel

                                    ┄┄┄┄┄ Signaling channel

                                    ───── Hand-over control channel

Figure 5.3 MCS-L1 network.

There are two types of control channels: a paging channel for control of mobile terminating call setup and an access channel for control of mobile originating call setup and location registration. Signals in the paging channel can be transmitted under the complete control of the base station. Accordingly, the paging channel can be used very efficiently, whereas the efficiency of the access channel is low because it is used in a random access scheme by mobile stations. Each control channel is used in a multitransmitter simulcasting scheme; that is, control signals are transmitted simultaneously from all the base stations in a location registration area on the same frequency. The simulcasting scheme has the merit of efficient wide-area paging, but it necessitates both baseband and carrier synchronization. Signal transmission quality is generally degraded in an overlapped area due to incomplete carrier synchronization. In MCS-L1, however, the quality is not degraded but is greatly improved, compared to that without overlap, by offsetting carrier frequencies of adjoining cells by 500 Hz, a separation greater than the baseband signal rate of 300 bps [8]. A two-to-one voice compandor (compressor and expander) is used to improve signal-to-noise ratio by compressing the variation of voice signal level. The automobile switching center (AMC) contains the home memory of each subscriber's data, including location registration.

MCS-L2 is NTT's second-generation analog cellular system and is designed for higher subscriber capacity. Its main features are listed in Table 5.3 [9].

Table 5.3
Main Features of MCS-L2

Item	*Feature*
Frequency band	800 MHz
Transmit and receive separation	55 MHz
Channel separation	12.5 kHz (6.25 kHz with interleaving)
Number of channels	2,400
Modulation	PM (maximum deviation 2.5 kHz) with 2:1 compandor
Signaling	2400 bps common control, and 100 bps channel-associated biphase
Control Channels	Paging channel with hybrid casting and access channel
Diversity	Postdetection selection

The network is similar to that of MCS-L1, except that the MCS and AMC in MCS-L1 are integrated as the mobile control center (MCC). Diversity reception is adopted in both mobile or portable units and base stations. The channel separation is narrowed to 12.5 kHz (6.25 kHz by interleaving), and the maximum frequency deviation is 2.5 kHz, one-half that of MCS-L1. The number of channels a mobile unit has is 2,400, because the number of channels doubles when the channel separation is halved. However, the average carrier-to-interference ratio (CIR) necessary for good speech quality increases by 2 to 3 dB in a Rayleigh fading environment. This increase results in lower efficiency in geographical frequency reuse. The overall effect of channel separation narrowing must be considered as a tradeoff between the increased number of channels and decreased geographical reuse efficiency, as shown in Figure 5.4.

In this figure, geographical traffic distribution is assumed to be the function of $10^{-r/k}$, where r (km) is the distance from the center of the city, and k (km) is the parameter of concentration, which is 10 to 20 km in the Tokyo area. The parameter ∂ is R/k, where R (km) is the cell radius. The figure shows that the spectrum utilization factor is maximum when the peak deviation is around 2 to 2.5 kHz with a cell radius of 1 to 5 km ($k = 10$ to 20 km and $\partial = 0.1$ to 0.25). Maximum deviation of 2.5 kHz corresponds to 12.5-kHz channel separation. This is the reason for selecting 12.5-kHz channel separation.

Postdetection selection diversity is employed in mobile and portable units and in base stations as well. Diversity improves the necessary CIR value by 5 to 6 dB and results in a 6- to 7-dB improvement in the necessary value of the average carrier-to-noise ratio (CNR) [10]. Interference detection and channel switching is also employed; here, the CIR value in the speech channel is measured, and if it exceeds a threshold, the call is switched to a channel with less interference in the same cell. When used on elevated freeways or in skyscrapers, mobile units often interfere considerably with the reception at base stations using the same frequency in a distant cell. To reduce such excess interference, adaptive power control is used in both base and mobile stations [11]. Adaptive transmitter power

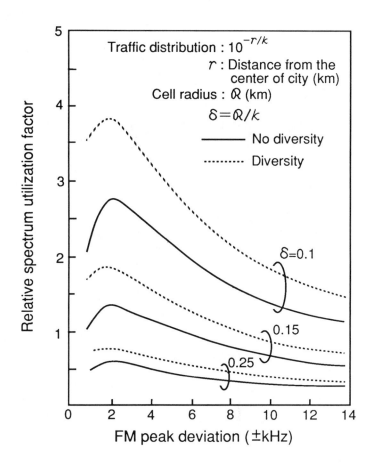

Figure 5.4 Effect of channel narrowing.

control minimizes the amount of transmitter power by sending data on the average reception level back to the transmitting side. This also reduces power consumption in portable units.

The data rate of common control channels is 2,400 bps. This rate makes the error rate improvement with multitransmitter simultaneous transmission ineffective because the necessary frequency offset is more than 1,500 Hz, which is too great to apply to a system with 2.5-kHz maximum frequency deviation. For inservice control signaling, both 100-bps and 2,400-bps signaling is used. The 100-bps channel-associated control signal is used for signals including power control and interference reporting, and it is transmitted during a call without interrupting it. Two types of common control channels are used: a paging channel and an access channel. A new transmission method called *hybrid casting* is used in the paging channel. In this method, signals common to all the cells in a control zone, such as subscriber paging, are simulcast, as shown in Figure 5.5. Signals peculiar

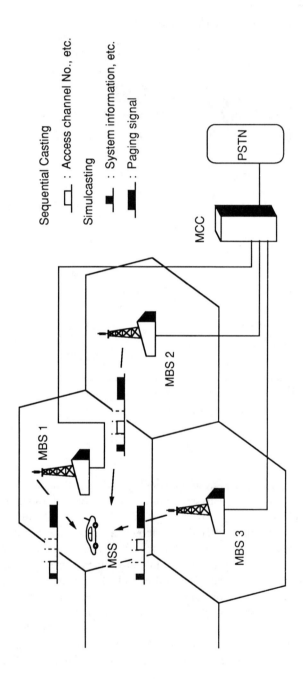

Figure 5.5 Basic concept of hybrid casting.

to each cell, such as each cell's access channel number, are broadcast sequentially. Mobile units select the cell they are in by detecting the highest level signal among sequentially broadcast ones. Hybrid casting allows both wide-area paging by simulcasting and small-area locating by sequential broadcasting. A dedicated access channel (or channels) is allocated to each base station and its channel number is broadcast by sequential signals of the paging channel.

Newly designed portable and mobile units can access either MCS-L1 or MCS-L2. This allows MCS-L2 to be deployed partially overlapped with the existing MCS-L1 areas, and also allows the MCS-L1 system to be gradually displaced by the more frequency-effective MCS-L2. So a rollout of the new network can be carried out gradually as demand dictates, thus spreading the capital outlay.

5.2.3 IDO and DDI Systems

As mentioned before, IDO uses both an MCS-L2-based system and NTACS and DDI uses JTACS and NTACS. The radio interface of the MCS-L2-based system is basically similar to that of MCS-L1 except for the frequency band and, accordingly, the number of channels. DDI's JTACS is basically similar to the U.K.'s TACS except for the frequency band and the number of channels. NTACS is a system developed by Motorola, Inc., and is basically similar to the U.S.'s NAMPS. NTACS doubles the system capacity by narrowing channel separation from 25 to 12.5 kHz. Channel-associated inservice control signaling is changed from 8 Kbps in JTACS to 100 bps in NTACS. NTACS and JTACS share common control channels. New mobile units are compatible with both JTACS and NTACS, as they are in the case of MCS-L1 and MCS-L2. The main features of JTACS and NTACS are summarized in Table 5.4.

Table 5.4
Main Features of JTACS and NTACS

	Feature	
Item	*JTACS*	*NTACS*
Frequency band	800 MHz	800 MHz
Transmit and receive separation	55 MHz	55 MHz
Channel Separation	50 kHz	25 kHz
With interleaving	25 kHz	12.5 kHz
Modulation	PM with 2:1 compandor	PM with 2:1 compandor
Maximim deviation	9.5 kHz	5 kHz
Signaling		
Common control	8 Kbps biphase	8 Kbps biphase
Channel-associated	8 Kbps biphase	100 bps biphase

5.3 OPERATIONAL EXPERIENCE IN JAPAN

5.3.1 History of Service and Operation

NTT began the first cellular service in December 1979 in the Tokyo metropolitan area with vehicle-mounted mobile units 6,600 cm^3 in volume and 7 kg in weight. The system was successively deployed in Osaka and other areas. The dialing code for calling the mobile unit was 030+CD+EFGHJ. The number 030 is the mobile service access number and CD was the code for an area in which the subscriber's home memory was registered. EFGHJ was the subscriber number. Until March 1984, the AMCs in each service area were not connected to each other, and roaming between service areas was not possible. In March 1984, nationwide roaming service began. At that time, however, the calling procedure for the roaming unit was not automatic but based on area designation dialing because of charging system restrictions of the fixed network. The calling party had to dial code CD for the area in which the called mobile unit was supposed to be. If the location registration of the called unit did not coincide with the dialed code, the system informed the calling party of the correct CD code in which the called unit had registered its location. Charging rate (i.e., the time in seconds for which every 10 yen is charged) was determined according to the distance between the calling party location and the location of the dialed CD area.

In January 1982, new 1500-cm^3, 2.4-kg mobile units were introduced. It was September 1985 when the transportable units were placed on the market. The units transmitted 5W, the same power as the vehicle-mounted units, were 2,300 cm^3 in volume and weighed 3 kg. In April 1987, portable units became available whose figures were 1W, 500 cm^3, and 750g, respectively.

The numbering capacity was initially 100,000 (five-digit subscriber number), and afterwards a new service access number, 040, was added to double the capacity. In March 1988, a new numbering and dialing system, area nondesignation dialing, was introduced [12]. In this system, 030 and 040 are codes to distinguish one of the two charging rates. The rates correspond to whether the distance between the calling and the called parties is within, or over, 160 km. The subscriber number became CDEFGHJ (i.e., a seven-digit number). The CD code also allowed discrimination of the carrier to which the subscriber is contracted. Therefore, the numbering capacity expanded to 10,000,000. The number 030 is for charging within a 160-km range, and 040 is for charging over 160 km. The calling party dials 030 plus the seven-digit subscriber number when the called party is supposed to be located within 160 km. When the called party is supposed to be more than 160 km, the caller dials 040. When 030 or 040 dialing does not correspond to the actual distance between the calling and the called parties, the system informs the caller to redial with the correct code 040 or 030.

In May 1988, MCS-L2 was introduced in the Tokyo metropolitan area to cope with the expanding demand, the annual growth of which exceeded 50%, and was successively deployed in other major metropolitan areas. Old-type mobile units, which could access only MCS-L1, were actively recalled and replaced with new dual-mode units. In October 1988, approximately nine years after the start of the service, the number of subscribers exceeded 100,000. The growth rate, however, was rather low compared to other countries.

Soon after the 100,000th subscriber came into the market, the era of NTT's monopoly ended, ushering in the age of competition. In December 1988, IDO started a service in the Tokyo metropolitan area with the system based on NTT's MCS-L2. In July 1989, the Kansai Cellular Telephone Company, a DDI subsidiary, began cellular service with JTACS in the Osaka metropolitan area, the largest city in western Japan. DDI's seven other subsidiaries have begun or plan to initiate service successively in other areas of Japan. The service areas of cellular operators are shown in Figure 5.6. The only nationwide system is NTT's MCS-L1 and MCS-L2. IDO and DDI subscribers can use the roaming service of MCS-L1 with mobile units compatible with MCS-L1 and the IDO or DDI systems. However, IDO's and DDI's roaming subscribers must also subscribe to the NTT network. In August 1991, a pair of 2-MHz blocks in the 800-MHz band were newly allocated to IDO and DDI. They started new services in this frequency band with NTACS within the year.

5.3.2 Subscriber Growth

The growth of the number of subscribers is shown in Figure 5.7. The number of subscribers in October 1988 exceeded 100,000, and in July 1991 it reached 1 million. The number of subscribers as of the end of September 1991 was 1,089,200. Of this number, 720,000 (66%) were portable subscribers. NTT had 668,000 (61%) customers. The figures for IDO and DDI were 193,000 (18%) and 228,000 (21%), respectively, as listed in Table 5.5. The Telecommunications Technology Council of the Ministry of Post and Telecommunications (MPT) reported in 1984 that the figure will be 4,500,000 for mobiles and 1.2 million for portables in the year 2000. Private gatherings, or study groups, of the ministry also issued figures: at least 8 million mobiles and portables in the year 2000.

The annual growth rate of the number of subscribers for the last several years was at least 50%. If the average growth rate is 50%, the number of subscribers will reach 40 million in the year 2000. Even if the growing rate reduces to 20% on average, the number of subscribers will be more than 5 million. Therefore, the number of subscribers in the year 2000 will exceed the numbers, 4.5 million mobiles and 1.2 million portables, which were predicted by the MPT Council.

5.3.3 Mobile and Portable Units

As mentioned earlier, the first mobile unit was 6,600 cm^3 in volume. In January 1982, it had shrunk to 1,500 cm^3. The first portable unit in the 500-cm^3-volume range was introduced by NTT in April 1987. In February of 1989, NTT introduced a 400-cm^3 dual-mode (MCS-L1 and MCS-L2) portable unit that has a postdetection diversity reception function. In

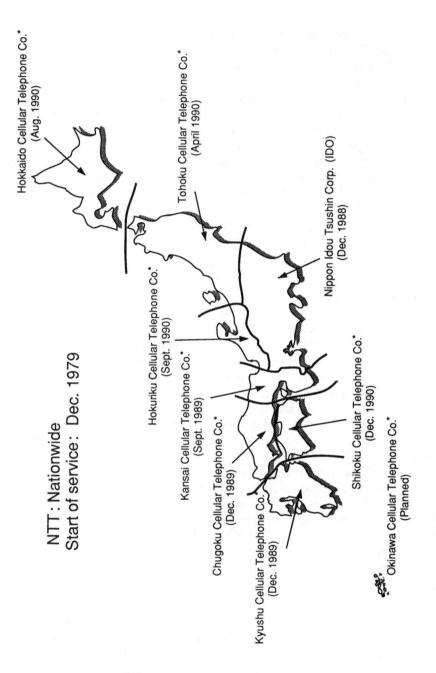

NTT : Nationwide
Start of service : Dec. 1979

Hokkaido Cellular Telephone Co.*
(Aug. 1990)

Tohoku Cellular Telephone Co.*
(April 1990)

Nippon Idou Tsushin Corp. (IDO)
(Dec. 1988)

Hokuriku Cellular Telephone Co.*
(Sept. 1990)

Kansai Cellular Telephone Co.*
(Sept. 1989)

Chugoku Cellular Telephone Co.*
(Dec. 1989)

Shikoku Cellular Telephone Co.*
(Dec. 1990)

Kyushu Cellular Telephone Co.*
(Dec. 1989)

Okinawa Cellular Telephone Co.*
(Planned)

Figure 5.6 Service areas of cellular carriers. * = DDI's subsidary company.

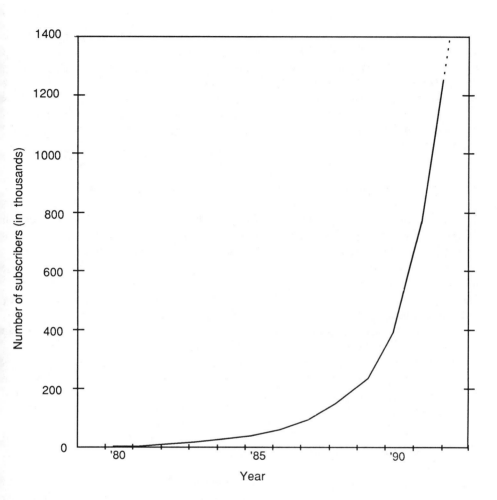

Figure 5.7 Subscriber growth.

Table 5.5
Number of Subscribers Between Carriers,
September 1991

Carrier	*Number of Subscribers (1000s)*	*Percentage*
NTT	668	61
IDO	193	18
DDI	228	21

October of that year, DDI introduced the "Handy-Phone," based on Motorola's Micro TAC portable. It is 221 cm³ in volume and weighs 303g, and it offers 30 minutes of talk time and 8 hours standby. This marked the beginning of the miniaturization war. In September 1990, IDO's "MINIMO" was put onto the market. It is 203 cm³ in volume and 298g in weight, and it offers 30 minutes of talk time and 9 hours standby. NTT's "Mova," introduced in May 1991, is 150 cm³ in volume and 230g in weight and offers 45 minutes of talk time and 6 to 13 hours of standby. IDO also began "Tokyo Phone," based on Motorola's Micro TAC, in October 1991.

5.3.4 Problem Areas

The most critical problem of Japanese analog cellular systems is a predicted shortage of subscriber capacity to meet the demand in the near future. The next most serious problem is that of nonstandardized systems, which results in limited and inconvenient roaming, low trunking efficiency caused by channel splitting between systems, and so on. As mentioned above, IDO or DDI customers who want to receive roaming service with NTT's MCS-L1 must subscribe to both systems. This means that they have to pay subscription fees to two carriers and that they need two mobile unit numbers. Another problem is poor transmission quality for nonvoice signals, including facsimile and modem signals. Voice privacy is also a problem. For subscribers who wish to prevent eavesdropping, a voice privacy service is provided at extra cost. However, voice privacy is still a problem because subscribers without the voice privacy service fear they may be the target of an eavesdropper, even though it is almost impossible to listen in on a specific conversation. These problems should be solved in the digital cellular system.

5.4 COMMERCIAL ISSUES

Mobile transceiver units are deemed part of the network, as well as base stations and switching stations in Japan. Handsets of mobile units or other terminals are subscriber equipment. Therefore, all mobile transceiver units and portables are the property of cellular carriers, and they are rented to subscribers. This helped the smooth transition from MCS-L1 to MCS-L2, as mentioned in 5.3.1. Carriers can delegate the role of subscription management and maintenance of mobile units to third parties.

The charges consist of the subscription fee, network access charges (including the renting of the mobile unit), and charges for calls. The calling party, either mobile or land subscriber, pays call charges, including air time charges. Call charges from NTT's fixed network to any cellular carrier's unit are the same. On the other hand, call charges from a cellular unit depend on the carrier it subscribes to. The charges and fees of each carrier are listed in Table 5.6. The access charge, including rent, for a cellular unit was 30,000 yen a month when NTT introduced the first system in 1979. It has decreased to 11,000 to 13,000 yen a month. Also, weekday busy-hour call charges, local and long distance,

Table 5.6
Fees and Charges (in Yen) of Japanese Cellular Services, February 1992

	Carrier		
	NTT	*IDO*	*KCT**
Initial fee	145,800†	53,900‡	43,800
Access Charge◇			
Mobile	13,000	11,000	11,000
Transportable	13,000	12,000	12,000
Portable	17,000 (mova)	15,000 (minimo)	15,000 (Micro TAC)
Call Charge§			
Mobile			
Within prefectures#			10/9.5 sec
Within 160 km	10/7 sec	10/8 sec	10/7.5 sec
Over 160 km	10/5.5 sec	10/6 sec	10/6 sec
Fixed telephone			
Within 160 km		10/7 sec	
Over 160 km		10/5.5 sec	

*Kansai Cellular Telephone Co., DDI's largest cellular subsidiary
Includes deposit of 100,000 yen which will be returned three years later. Does not include peripheral equipment, such as a battery charger.
Includes insurance amount of 8,100 yen
◇Monthly charge including mobile unit rent fee.
Weekday busy hours.
#KCT service area prefectures and contiguous prefectures to the one in which originating mobile unit locates.

were 10 yen for 6.5 seconds and 10 yen for 2.5 seconds, respectively. These have also decreased to 10 yen for 7 to 9 seconds and 10 yen for 5.5 to 6 seconds.

REFERENCES

[1] Watanabe, M., and K. Miyauchi, "Research and Development on Land Mobile Telephone Systems," *Review of the ECL*, Vol. 25, Nos. 11 and 12, 1977, pp. 1141–1146.
[2] Okumura, Y., et al., "Field Strength and Its Variability in VHF and UHF Land Mobile Service," *Review of the ECL*, Vol. 16, Nos. 9 and 10, 1968, pp. 825–873.
[3] Jakes, W. C., ed., *Microwave Mobile Communications*, John Wiley & Sons, 1974, pp. 90–119.
[4] Hata, M., "Empirical Formula for Propagation Loss in Land Mobile Radio Services," *IEEE Trans. on VT*, Vol. VT-29, No. 3, 1980, pp. 317–325.
[5] Araki, K., "Fundamental Problems of Nationwide Mobile Radio Telephone System," *ECL Technical Journal*, NTT, Vol. 16, No. 5, 1967, pp. 843–865 (in Japanese).
[6] Sakamoto, M., and M. Hata, "Effects of Channel Narrowing in Small Zone FM Mobile Communication Systems," *IEEE Trans. on VT*, Vol. VT-36, No. 1, 1987, pp. 14–18.
[7] Kamata, T., M. Sakamoto, and K. Fukuzumi, "800 MHz Band Land Mobile Telephone Radio System," *Review of the ECL*, Vol. 25, Nos. 11 and 12, 1977, pp. 1157–1171.
[8] Hattori, T., and K. Hirade, "Multi Transmitter Digital Signal Transmission by Using Offset Frequency Strategy in a Land-Mobile Telephone System," *IEEE Trans. on VT*, Vol. VT-27, No. 4, 1978, pp. 231–238.
[9] Kuramoto, M., K. Hirade, and M. Sakamoto, "Design Concept of New High-Capacity Land Mobile Communication System," *Proc. ICC'84*, 1984, pp. 1188–1191.

[10] Hattori, T., "Major Techniques for High-Capacity Land Mobile Communication System—Voice and Data Transmission," *NTT International Symposium*, 1983, pp. 44–63.
[11] Sakamoto, M., and M. Hata, "Efficient Frequency Utilization Techniques for High-Capacity Land Mobile Communications System," *Review of the ECL*, Vol. 35, No. 2, 1987, pp. 89–94.
[12] Nakajima, A., and K. Yamamoto, "Advanced Mobile Communication Network Based on Signaling System No. 7," *Proc. ICC'87*, 1987, pp. 747–752.

Part III
Digital Cellular Radio Systems

Chapter 6
The Pan-European System: GSM

D. M. Balston
Intercai Mondiale, Ltd.

During 1992 and 1993, the countries of Western Europe have been introducing a new cellular radio system to replace the many incompatible standards presently in service. The initiative provides the European mobile communications industry with a home market of 300 million people, while at the same time providing it with a significant technical challenge. This chapter discusses the background of the standard, outlines the system, and discusses its deployment and evolution. Chapter 7 describes some of the technical problems faced by the product development teams.

6.1 INTRODUCTION

As the European community starts to dismantle its trade barriers, the mobile communications field anticipated the trend with the creation of a pan-European mobile telephone system. For the first time in the history of the European electronics industry, political, commercial, and industrial forces have come together to generate a home market for European industry matching that of the United States.

That this initiative is in the rapidly expanding and technologically challenging field of cellular radio makes the opportunity all the more exciting. This enterprise offers European industry the chance to participate in equipment supply in a field where two U.K. cellular operators, Cellnet and Vodafone, have shown the most rapid growth worldwide.

At present, there are six analog systems operational in Europe, and a mobile designed for one cannot be used with another. Chapters 3 and 4 describe the two major standards in operation in Europe, but in addition to these there are C450 in Germany, RadioCom 2000 in France, and RTMS in Italy. None of these systems interwork.

In 1981, a joint Franco-German study was initiated to develop a common approach that, it was hoped, would become a standard for Europe. Soon after, the main governing body of the European PTTs (CEPT) set up a committee known as the Groupe Special Mobile (GSM) under the auspices of its Committee on Harmonization (CCH). GSM was charged with defining a system that could be introduced across Europe in the 1990s. The stage has now been reached where equipment has been installed and commissioned, services are being brought into operation in several European countries, and the acronym has been updated to refer to the "Global System for Mobile Communications."

6.2 CREATING THE SPECIFICATIONS

The early years of GSM were devoted primarily to the selection of the radio techniques for the air interface. By 1986, GSM was ready to undertake trials of the different candidate systems proposed for this interface, and later that year six different systems underwent trials in Paris. GSM had drawn up a list of seven criteria ranked in order of importance to be used in assessing these candidates. At the top of the list was spectral efficiency, measured as the number of simultaneous conversations per megahertz per square kilometer. Some of the criteria to be met by the GSM system include:

- Spectral efficiency,
- Subjective voice quality,
- Mobile cost,
- Hand-portable feasibility,
- Base station costs,
- Ability to support new services, and
- Coexistence with existing systems.

At the same time, the United Kingdom, which did not have a candidate in the field, undertook a study of the different systems against the criteria listed above and started the development of a test-bed to assess the critical features. None of the candidates was selected, but the information gleaned was used to generate the outline specifications of a system that capitalized on ideas and approaches from several of the candidates.

On one subject there was no serious debate. The performance of cellular radio is restricted primarily by cochannel interference, and a given quality of telephony can be achieved at much higher levels of cochannel interference if digital transmission is used. This allows the cells to be reused more frequently, and it has been estimated that this factor alone offers the GSM system a three-fold improvement in spectral efficiency over the baseline NMT system.

Another factor that encouraged the trend towards digital transmission was that the telecommunications industry worldwide has been converting rapidly to digital methods, and the advent of ISDN will demand a much higher level of digital signaling than has hitherto been the case [1].

Equally important is the evolutionary path offered by digital transmission. If the system is to stand the test of time, it is essential that it is able to evolve to accommodate improvements that future competing systems might introduce. The most important of these currently foreseen is the introduction of an 8-Kbps speech coder. A digital system is more easily configured to change channel characteristics than is an analog system.

Considerable debate did take place, however, over the most suitable transmission method. Various approaches were represented by the candidate systems (FDMA, TDMA, CDMA), and the final decision to adopt a TDMA structure was made in April 1987. By that time, the structure of the GSM specification team had grown substantially and had practically reached its final form, shown in Figure 6.1. In 1989, responsibility for the specification passed from CEPT to the newly formed European Telecommunications Standards Institute (ETSI).

The period 1987 to 1990 saw a tremendous effort by the working parties and their supporting expert groups to create and document a complete mobile telecommunications system. The specifications and their explanatory notes were substantially completed by 1990 and run to 138 documents, some of which are several hundred pages long. They are divided into 12 sets of recommendations [2] covering different aspects of the system, as shown in Table 6.1. These make extensive reference to relevant CCITT, CEPT, and International Standards Organization (ISO) standards, and provide system definitions, standardization aspects, mandatory and optional features, and guidance on system design.

6.3 SYSTEM DESCRIPTION

6.3.1 Open Interfaces

The main philosophy underlying the GSM approach is that the major interfaces in the system should be open and in the public domain. This has the twin benefits of allowing supplier competition for each network node and encouraging the evolution of ideas, albeit within the constraints of the interface definitions. Considerable time and energy have been dedicated to the development and standardization of all the interfaces shown in bold in Figure 6.2. In terms of the standard ISO seven-layer model, GSM specifies layers 1, 2, and 3 (the physical, data link, and network layers) [3]. This approach represents a significant departure from previous practice, and it is one of the major aspects in which the GSM system differs from existing systems.

6.3.2 System Components

In a chapter such as this, it is only possible to give an overview of the major system elements and briefly describe their functions. In the following, each of the system components shown in Figure 6.2 will be described as it comes into operation when a mobile station makes a call. The acronyms used by GSM and reflected in Figure 6.2 are defined in Table 6.2.

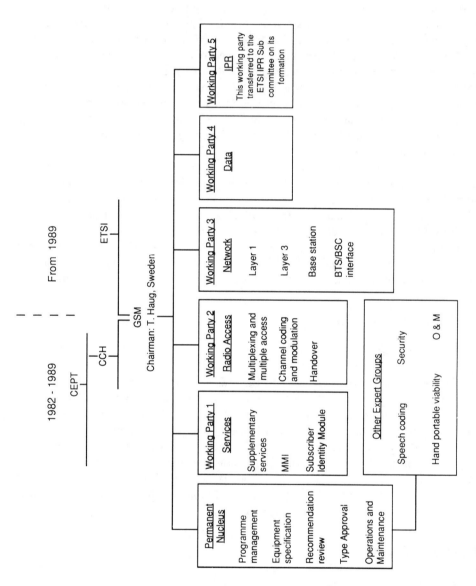

Figure 6.1 The GSM committee structure.

Table 6.1
GSM Recommendations

Number	Subject
00	Preamble
01	General
02	Service aspects
03	Network aspects
04	MS-BS interface and protocols
05	Physical layer on the radio path
06	Audio aspects
07	Terminal adapters for mobile stations
08	BTS/BSC and BSC/MSC interfaces*
09	Network interworking
10	Service interworking†
11	Equipment specification and type approval specification
12	Operations and maintenance

* A list of GSM acronyms and abbreviations is given in Table 6.2 and included in the general table of acronyms at the end of the book.

† Planned but never implemented.

6.3.2.1 Mobile Station

The mobile station comes in a number of different forms, ranging from the traditional car-mounted phone operating at 20W, to transportables operating at 8W and 5W, to the increasingly popular hand-portable units, which typically radiate less than 2W. A fifth class for hand-portables operating at 0.8W has been specified for microcellular versions of the network.

One of the main factors governing the hand-portable size and weight is the battery pack. Several features of the system are designed to allow this either to be smaller or to give a substantially longer life between charges. Chief among these is discontinuous receive (DRX). This allows the mobile to synchronize its listening period to a known paging cycle of the network. This can typically reduce the standby power requirements by 90%.

6.3.2.2 Radio Subsystem

When the mobile user initiates a call, his or her equipment will search for a local base station. A base station subsystem (BSS) comprises a base station controller (BSC) and several base transceiver stations (BTS), each of which provides a radio cell of one or more channels. The BTS is responsible for providing layers 1 and 2 of the radio interface (i.e., an error-corrected data path). Each BTS has at least one of its radio channels assigned to carry control signals in addition to traffic. The BSC is responsible for the management

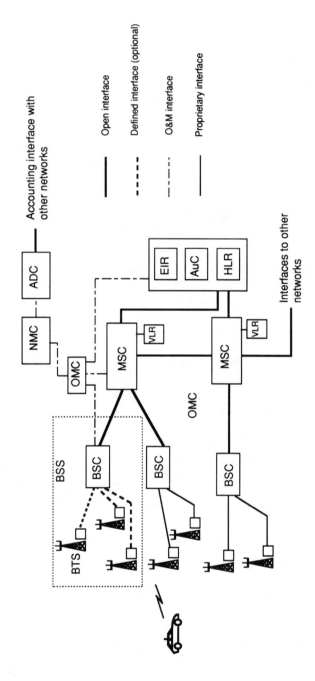

Figure 6.2 GSM network components and standardized interfaces.

Table 6.2
GSM Acronyms

Acronym	Meaning
MS	Mobile station
BTS	Base transceiver station
BSC	Base station controller
BSS	Base station subsystem
MSC	Mobile services switching center
HLR	Home location register
VLR	Visitor location register
AuC	Authentication center
EIR	Equipment identity register
OMC	Operations and maintenance center
NMC	Network management center
ADM	Administration center

of the radio resource within a region. Its main functions are to allocate and control traffic channels, control frequency hopping, undertake handovers (except to cells outside its region), and provide radio performance measurements. Once the mobile has accessed and synchronized with a BTS, the BSC will allocate to it a dedicated bidirectional signaling channel and will set up a route to the MSC.

6.3.2.3 Switching Subsystem

The MSC routes traffic and signaling within the network and interworks with other networks. It comprises a trunk ISDN exchange with additional functionality and interfaces to support the mobile application. When a mobile requests access to the system, it has to supply its international mobile subscriber identity (IMSI). This is a unique number which will allow the system to initiate a process to confirm that the subscriber is allowed to access it. This process is called *authentication*. Before it can do this, however, it has to find where the subscriber is based. Every subscriber is allocated to a home network associated with an MSC within that network. This is achieved by making an entry in the HLR, which contains information about the services the subscriber is allowed. The HLR also contains a unique authentication key and associated challenge/response generators.

6.3.2.4 Mobility Management and Security

Whenever a mobile is switched on, and at intervals thereafter, it will register with the system; this allows its location in the network to be established and its location area to be updated in the HLR. A location area is a geographically defined group of cells. On first registering, the local MSC will use the IMSI to interrogate the subscriber's HLR and will add the subscriber data to its associated VLR. The VLR now contains the address of the subscriber's HLR, and the authentication request is routed back through the HLR to the subscriber's authentication center (AuC). This generates a challenge-response pair that is used by the local network to challenge the mobile. In addition, some operators also plan to check the mobile equipment against an EIR in order to control stolen, fraudulent, or faulty equipment.

The authentication process is very powerful and is based on advanced cryptographic principles. It especially protects the network operators from fraudulent use of their services. It does not, however, protect the user from eavesdropping. The TDMA nature of GSM, coupled with its frequency hopping facility, will make it very difficult for an eavesdropper to lock onto the correct signal, however, and there is thus a much higher degree of inherent security in the system than is found in today's analog systems. Nevertheless, for users who need assurance of a secure transmission, GSM offers encryption over the air interface. This is based on a public key encryption principle and provides very high security.

6.3.2.5 Call Setup

Once the user and his or her equipment are accepted by the network, the mobile must define the type of service it requires (voice, data, supplementary services, etc.) and the destination number. At this point, a traffic channel with the relevant capacity will be allocated and the MSC will route the call to the destination. Note that the network may delay assigning the traffic channel until the connection is made with the called number. This is known as *off-air call setup*, and it can reduce the radio channel occupancy of any one call, thus increasing the system traffic capacity.

6.3.2.6 Handover

GSM employs mobile-assisted handover. In this technique, the mobile continuously monitors other base stations in its vicinity, measuring signal strength and error rate. These measurements are combined into a single function and the identities of the best six base stations are transmitted back to the system. The network can then decide when to initiate handover. The use of bit error rate, in addition to signal strength, adds considerably to the ability of the network to make informed handover decisions and is another example of the advantage of digital transmission over analog. The BSC can initiate and execute handover if both BTSs are under its own control. In this instance, the BSC can be considered the manager of a specific group of radio frequencies for a geographic region and can control that resource to maximize its utilization. Alternatively, and whenever handover must take place to a cell outside the control of the BSC, the MSC controls and executes handover.

6.3.2.7 *Mobile-Terminated Calls*

When setting up a call from the fixed network to the mobile, the procedure is much the same. First, however, the mobile must be found. This is achieved by means of a paging signal that covers the location area in which the mobile has registered. Mobiles continuously monitor the paging channel, and on detecting a call to them, undertake the access procedure described. Note that authentication is not mandatory for received calls, but it is expected that most operators will demand it.

The paging procedure has been designed to facilitate significant battery-saving potential in the hand-portable. Unless a hand-portable is used excessively, the biggest drain on its battery comes not from the time spent using it, but from the standby cycle as it monitors the paging channel in case it is being called. In the GSM system, the DRX mode (discussed in Section 6.3.2.1) allows the mobile, once it has located the paging signal, to synchronize a clock, knowing that it will not get another signal until a specified time has elapsed. It can thus power down its circuits for most of the time during standby.

6.3.3 Network Management

So far we have discussed the network elements used directly when making a call. A network of this complexity, however, needs to be managed and maintained. This is the function of the OMC. GSM leaves decisions on operations and maintenance (O&M) to the individual operators, but general guidelines are given that reinforce the move towards an overlaid telecommunications management network. In this approach, five separate management functions are identified:

- Operations and performance management,
- Maintenance,
- System change control,
- Security management, and
- Administration and commercial functionality.

The activities covered by each major area are further defined in Table 6.3.

The telecommunications management network is hierarchical in structure. It starts with the O&M functionality of individual network elements, integrates them with OMCs and provides a network management center (NMC), which gives an overall picture of the operation of the PLMN.

6.4 AIR INTERFACE

6.4.1 Frequency Allocation

Throughout Europe, GSM has been allocated a specific 50 MHz of spectrum divided into transmit and receive bands separated as shown in Figure 6.3. It should be noted, however,

Table 6.3
Activities of Network Management Functions

Operations and Performance Management

Network Status Information	Operations	Performance Data	Performance Management
Circuits	Interworking	Traffic measurements	Radio network
Nodes	Network configuration	Quality of service	management
Signaling	Control of system	observations	Switching network
	elements	Availability	management
	Node configuration	performance data	Routing control
		Throughput	
		measurements	
		Handover statistics	

Maintenance

BSS	MSC	Transmission	Database
BTS	Hardware	Lines	HLR
BSC	Software	Microwave	AuC
Mobile integrity		Multiplexing	EIR

System Change Control

Enhancements	Extensions	Reconfigurations
Provision of new features and new functions	Addition of existing equipment and existing services	Reorganization of existing equipment

Security Management

Network Access	Security Reporting	Data Security Management
Management of authorization, access control, authentication facilities	Reporting on feature status access status Management status	Management of keys and encryption Security routing Subscriber identity module

Administrative and Commercial

Billing	Customer Inquiries
Tariff Charging and accounting	Subscriber and mobile-equipment management Customer services

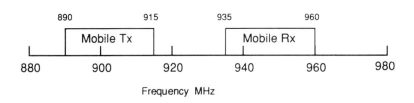

Figure 6.3 GSM spectrum allocation.

that not all countries are able to use the full allocation at present. This is due to existing commitments, often military in nature. Each radio channel is 200 kHz wide, and thus there are a total of 125 paired channels available. In practice, where there is more than one operator, guard bands have to be provided between the frequency bands allocated to them, and thus the actual number of channels available is less than the maximum. The salient features of the air interface are shown in Table 6.4.

6.4.2 Speech Coding

The spectral efficiency target set for the GSM system demands a speech codec that can provide toll-quality speech at 16 Kbps or less. The solution adopted is based on a residually excited linear predictive coder (RELP) enhanced by the inclusion of a long-term predictor (LTP) [4]. This improves the speech quality by removing the structure from the vowel sounds prior to coding the residual data. It has the effect of removing the coarseness often associated with linear predictive coding, especially on female voices. The basic data rate from the coder is 13 Kbps, and speech is processed in 20-ms blocks, as shown in Figure 6.4.

The resulting code is split into two parts, the most critical bits being put first. This first part has a half-rate convolutional code applied to it, and when recombined with the second part, the total block length is 456 bits. As we will see later, this block length can

Table 6.4
Basic Air Interface Parameters

Feature	Parameter
Channel spacing	200 kHz
Modulation	GMSK
Modulation depth	B·T = 0.3
Data transmission rate	270.833 Kbps
Number of channels/band	8 (16)
User data rate (nominal)	16 (8) Kbps
TDMA frame period	4.62 ms
Time-slot duration	0.58 ms

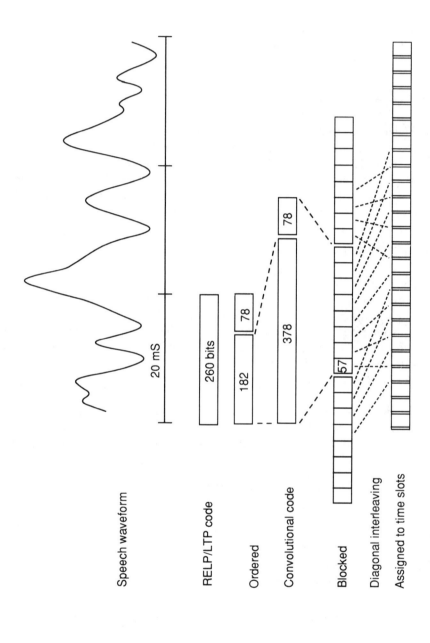

Figure 6.4 The speech coding process.

be fitted into four time slots, but in practice it is spread over eight. This process is called *diagonal interleaving* and it allows the convolutional code more chance to recover if a sequence of TDMA frames is badly corrupted during radio transmission.

6.4.3 Channel Structure

The specific parameters selected for the GSM air interface are shown in Table 6.4. The data transmission rate has been set as high as possible, commensurate with the cost-effective equalization of the expected multipath effects. The precise figure was then chosen to allow major system clocks to be derived from a common source.

The basic traffic data rate allows eight channels to be accommodated on a single RF carrier. With an eye to the future, however, the specification allows two separate channels to be interleaved within the same frame. This facility will effectively double the traffic capacity when a half-rate speech codec becomes available. Figure 6.5 shows the basic frame structure and the time slot organization for a traffic or signaling channel.

The 26-frame multiframe provides 24 frames allocated to traffic and 2 (the 13th and the last) allocated to control and supervisory signals associated with the traffic channels. Each traffic channel is assigned one of the 16 time slots in these 2 frames. (For 13-Kbps speech, where only eight traffic channels can be carried, the last TDMA frame is not used.) In addition to these two slow associated control channels (SACCH), the system provides:

- Fast associated control channels (FACCH), which steal time slots from the traffic allocation and are used for irregular control requirements such as handover;
- Stand-alone dedicated control channels (SDCCH), which are multiplexed onto a standard traffic channel and are used for registration, location updating, authentication, and call setup;
- Broadcast control channels (BCCH) (downlink only), which provide mobiles with base station identity and information pertaining to the cell;
- Random-access channel (RACH) (uplink only), which is used by the mobile to request access to the network;
- Access-grant channel (AGCH) (downlink only), which replies to a random access and assigns a DCH for subsequent signaling;
- Paging channel (PCH) (downlink only), which informs the mobile that the network needs to signal it.

Apart from the RACH, all these control channels have the same structure as the traffic channel shown in Figure 6.4. The RACH has a different equalizer training sequence because no system timing information is available at this stage (see Section 6.4.4).

Each time slot lasts 0.577 ms and comprises 148 bits, with an 8.25-bit guard period between slots. The traffic carried by the slot is divided into two separate 57-bit blocks and each block is assigned data from separate speech coding frames. Thus, eight such slots are needed to convey the 20 ms of speech data, but each slot is actually carrying

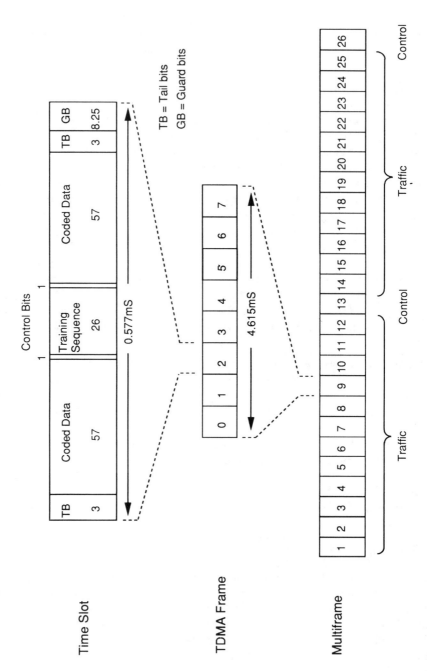

Figure 6.5 GSM frame structure.

data from two speech blocks simultaneously. Thus, four time slots in consecutive frames provide 456 traffic bits in 18.5 ms. This accommodates the 456 speech bits created every 20 ms, and the additional 1.5 ms adds up over 26 frames to provide the additional two control frames in the multiframe.

One control bit associated with each data block is used to flag whether the block is carrying normal traffic or has been stolen by the FACCH. The other bit denotes full- or half-rate speed. In the center of each slot is a sequence of 26 bits that are used by the receiver to set the parameters of equalizer and demodulator. This device is necessary to overcome the multipath problem described in Section 6.4.6.

6.4.4 Timing Advance

TDMA requires that the signals from all the mobiles using a single channel must reach the base station at the right time. They must not overlap each other. If the base station provides a reference signal, those mobiles nearest it will respond earlier than those nearer the perimeter of the cell. GSM has been specified to allow cells to extend up to 35 km from the base station. The time taken for a radio signal to travel the 70 km to the perimeter and back is 0.23 ms, and thus a guard period of this length would have to be provided on each time slot. This is clearly inefficient, and GSM overcomes this by informing the mobile how much in advance of the reference it should transmit in order to be correctly synchronized at the base station. This allows the guard period to be reduced to 0.03 ms (8.25 bits).

6.4.5 Modulation

The modulation scheme chosen by GSM is Gaussian minimum-shift keying (GMSK) with a bandwidth data-rate product (B·T) of 0.3 [5]. GMSK modulation produces a better defined spectral occupancy than FSK or differential phase modulation (DPM), and its resilience to cochannel interference, while not as good as that of DPM, is adequate for the GSM requirements.

6.4.6 Multipath and Equalization

At the frequency band occupied by GSM, radio waves do not refract very well and there are thus potentially many shadow areas created when a mobile or a base station transmits. This is compensated for by the tendency of the signals to reflect from buildings, hills, high-sided vehicles, and so on, and these reflections help to fill in the shadows. Many different reflections can reach the same point, however, and even when there is a direct path, it is not unknown for strong reflections to be received as well. The radio paths taken by the reflections must, of course, be longer than the direct path, and at the bit rates chosen for GSM, the difference in path length can be equivalent to several bit periods. Figure 6.6 demonstrates this effect, and it can be seen that the combined signal received at the mobile's antenna can be severely corrupted.

To date, radio systems have substantially avoided this multipath effect by choosing bit rates that are long compared with the expected multipath delays. Equalization is a technique, however, that allows us to recover the wanted signal despite severe multipath corruption. Until recently the cost and complexity of applying equalization was prohibitive. Now, however, 50,000-gate very-large-scale integration (VLSI) devices are not unusual, and equalization can be reduced to a single, albeit complex, chip. This means that higher data rates can be used with consequent improvements in spectral efficiency.

Equalization works by making an estimate of the impulse response of the transmission medium and then constructing an inverse filter through which the received signal is passed. There are several methods for estimating the transfer function of the transmission path and several variations of algorithm associated with each; but whatever the method they all rely on receiving a known sequence of data. This is the training sequence, which is transmitted in the middle of each time slot. The receiver detects this sequence and, knowing what bit pattern it represents, is able to estimate the transfer function most likely to have produced the signal received. The calculation of the coefficients of the filter required to compensate for the response is then relatively straightforward.

The multipath effects can change very rapidly in practice. The wavelength at 900 MHz is only 30 cm, and thus a change in the differential path length of only 15 cm between two signals received at an antenna can change their interference from constructive to destructive. The GSM specifications are designed to accommodate vehicles moving at up to 250 kph, and thus the mobile could have moved up to 32 cm in the 46 ms between successive traffic channel time slots. Add to this the problems of reflections from other moving vehicles and it is clear that each time slot has to be treated independently. It is also important to provide the best possible estimate of the path characteristics, and it can be seen that placing the training sequence in the middle of the slot reduces the time between it and the data bit most distant from it.

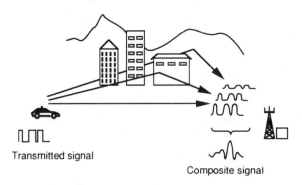

Transmitted signal

Composite signal

Figure 6.6 The multipath problem.

6.5 NETWORK FEATURES

6.5.1 GSM as an Intelligent Network

6.5.1.1 Intelligent Network Architecture

An intelligent network (IN) is identified as a network architecture that relocates specific services and databases from switches to one or more network control and decision points. The principal driving force behind INs is the inability of current network architectures to support the rapid development and deployment of "advanced" services due to the need to specify, develop, test, and deploy software in each switch in a public network, and to equip each switch with the subscriber/service data required for each new service. The move towards IN results in networks containing:

- Switches with bearer and basic service control capability known as service switching points (SSP);
- Elements with "advanced" service control capability known as service control points (SCP);
- A service management system (SMS) that controls the deployment of services and the associated service data.

The link between the SSPs (the switches) and their associated SCPs (the service logic/databases) will use the transaction capability (TC) part of SS7, with SCCP and MTP providing the basic mechanism on which to build the query portion of these new services, although for SCP functions collocated with an SSP, a local high-speed link may be substituted.

6.5.1.2 GSM Network Architecture

GSM has been designed with close reference to the IN model and can be considered to be the first true instance of an IN. It exhibits:

- An open distributed architecture;
- Separation of service control and switching functions;
- Full use of SS7 as the signaling communications infrastructure;
- Clearly defined and specified interfaces;
- IN structure.

With particular reference to the last point, examination of GSM network architecture in terms of SSP and SCP units indicates close alignment with the general IN architecture. This is illustrated in Figure 6.7.

The SSP is responsible for the interface with the service users. It provides a number of service components. The SSP also provides the bearer capability for telecommunications traffic and generates service triggers that cause service control requests to be directed at

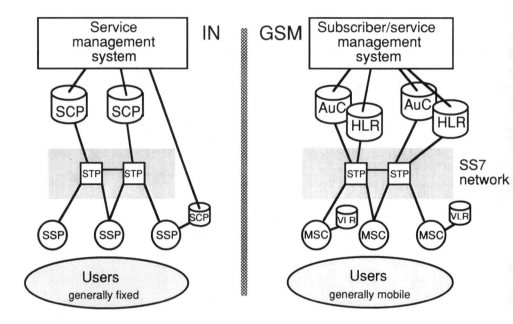

Figure 6.7 GSM and IN physical architectures.

the SCP, which may be remotely located for infrequently used services, or connected locally for frequently used services. In GSM terms, the SSP may be equated to the MSC and the associated radio subsystem responsible for providing service access to mobiles.

The SCP manages service triggers and controls the progress of a call based on the nature of the trigger and the service programs running on its internal processing platform. Referring to Figure 6.7 again, the SCP functionality may be equated with that of the HLR and other GSM database entities, such as the AuC and the VLR.

6.5.2 Services

GSM offers a wide range of services to its users. Not all these services will be available from day one, however, and the group formed to coordinate the views and commercial positioning of the operators (GSM Memorandum of Understanding (MoU)) has defined targets for the introduction of specific services. The main reasons for this are to provide guidance on priorities to the system suppliers and to ensure as far as possible that when subscribers roam they will be able to receive the same set of services as they are used to on their home network.

The GSM MoU has defined four categories for the introduction of service; three of these, E1, E2, and Eh are time-related; the fourth, A, means the introduction is optional at the operator's discretion. The time-related categories are:

- E1, start of service
- E2, end of 1994
- Eh, on availability of half-rate channels
- A, optional

The asynchronous (transparent and nontransparent) data services at rates at and above 2.4 Kbps are assigned E2 and subsequently Eh. All other bearer services are assigned category A. Teleservices are assigned categories as shown in Table 6.5. Supplementary services are assigned categories as shown in Table 6.6.

6.5.3 Subscriber Identity Module

An important innovation introduced by the GSM committee is the idea of using a *Smart Card* in conjunction with a mobile phone. The size of a credit card and containing a microprocessor and a small amount of memory, this card is being introduced in a number of fields, such as banking and security. Associating a cellular subscription with a card instead of a mobile phone introduces considerable flexibility. This is the concept of the subscriber identity module (SIM). It means that the subscriber can use any phone to receive incoming calls and have any outgoing calls charged to his or her own account. In effect, while the card is in the phone, it becomes a personal phone. All the subscriber's personal data, such as short-code dialing, services subscribed to, authentication key, and IMSI are stored in the smart card.

The downside of the approach is, of course, that every mobile phone has to have a card reader incorporated in it and the subscriber must remember to keep the card with him or her. GSM overcomes this disadvantage by allowing manufacturers to offer a semipermanent SIM that is plugged inside the equipment. This module is identical to the standard SIM except that it is mounted on a cut-down card with modified connectors. It is expected that this option will be particularly attractive to the hand-portable manufacturers.

Security-conscious subscribers will also want to incorporate a personal identity number (PIN) in their SIM so that nobody else could use it without their authority. GSM

Table 6.5
Introduction of Teleservices

Service	Introduction
Telephony	E1 then Eh
Emergency Calls	E1 then Eh
Transparent fax	E2
SMS Point to point, mobile terminated	E2
SMS Point to point, mobile originated	A
Nontransparent fax	A
SMS cell broadcast	A

Note: Reprinted with the kind permission of the GSM MoU Group.

Table 6.6

Introduction of Supplementary Services

Code	Supplementary Service	Introduction
CLIP	Calling line identification presentation	A
CLIR	Calling line identification restriction	A
CoLP	Connected line identification presentation	A
CoLR	Connected line identification restriction	A
CFU	Call forward unconditional	E1
CFB	Call forward on busy	E1
CFNRy	Call forward on no reply	E1
CFNRc	Call forward on not reachable	E1
CW	Call waiting	E2
HOLD	Call hold	E2
MPTY	Multiparty	E2
CUG	Closed user group	A
AoC	Advice of charge	E2
BAOC	Barring of all outgoing calls	E1
BOIC	Barring of outgoing international calls	E1
BOIC-exHC	Barring of outgoing international calls except to home country	A
BAIC	Barring of all incoming calls	E1
BIC-Roam	Barring of incoming calls when roaming	A

Note: Reprinted with the kind permission of the GSM MoU Group.

offers this facility. Indeed, the PIN is standard, but it is expected that most operators will choose to allow the user to disable the function after the first registration, although they will highly recommend its continued use. If it is not disabled, the PIN has to be entered every time the card is inserted in the phone and/or when it is switched on.

The introduction of a second PIN (PIN2) is planned to extend the use of some services, such as AoC, in phase 2.

6.5.4 Short-Message Service

A novel feature of GSM is the ability to transmit short data messages, each up to 160 alphanumeric characters long, over the signaling channels. This connectionless service operates rather like a pager function with the added advantages that the messages can pass in either direction and confirmation is provided that the message has been received.

Two types of short-message service (SMS) are planned, cell broadcast and point-to-point. In the cell broadcast case, a message is transmitted to all those active mobiles present in a cell that have the capability to receive SMS. This service is clearly only one-way, and no confirmation of receipt is obtained. It is expected that cell broadcast, which will be the first service to be offered, will be used to transmit messages on topics such as traffic conditions, sports results, and stock exchange information. The subscriber can arrange to filter out unwanted cell broadcast message types. The point-to-point service, as its name implies, allows a message to be sent to a particular mobile or lets the mobile send a message to a specific addressee in a service center.

Mobiles used for SMS services must contain special software to enable the messages to be decoded and stored. Typically, messages will be stored on the SIM and read using the standard display of the mobile.

The SMS requires a service center to receive incoming messages, check and organize them, send them to the operator, and receive and pass on any confirmation message. The operators themselves will no doubt set up their own service centers and possibly offer service providers the opportunity to connect their own service centers to the GSM network. In most cases, the subscribers will have to register with the SMS service provider and pay a monthly fee to receive the service.

6.5.5 Data

GSM recognized that data over cellular is going to play an increasingly important role in the future, and the digital nature of the standard makes it well suited to supporting such a service. It was perhaps one of the few failings of the committee, however, that the specification for data services encompasses all the existing services offered by the PSTNs throughout Europe. It was not possible, it seems, for the committee to cut through the variety and provide a service geared to the 1990s. GSM thus offers synchronous and asynchronous service, transparent and nontransparent, at data rates of 300 bps, 1.2 Kbps, 2.4 Kbps, 4.8 Kbps, and 9.6 Kbps. It will also support group 3 facsimile.

Chapters 13 and 14 describe in detail the data and facsimile services that are offered by GSM.

6.6 CURRENT STATUS

By September 1992, 27 operators representing 18 countries had committed themselves to GSM and joined the MoU Group. These are shown in Table 6.7.

Of these, Denmark, Finland, France, Germany, Italy, Norway, Sweden, and the United Kingdom already had one or more networks in operation and offering service before the end of 1992. The others expect to become operational during 1993.

The two networks in Germany and Vodafone in the United Kingdom appear to be undertaking the most rapid rollout of the network. There are reasons for this. In Germany there has been a very high level of suppressed demand for cellular services, and with two competing operators there is a race to attract the bulk of the unsatisfied customers. In the United Kingdom, on the other hand, Vodafone sees GSM as the vehicle to enable it to compete with the newly licensed PCN operators (see Chapter 8). GSM facilitates the deployment of microcells and allows service differentiation to be achieved, even between subscribers using the same base station. Vodafone have announced their Micro Cellular

Table 6.7
GSM Operators, September 1992

Country	Operator
Australia	TELECOM Australia
Austria	PTV Austria
Belgium	RTT Belgacom Belgium
Denmark	TELE Danmark Mobile, Dansk Mobil Telefon DMT
Finland	Telecom Finland, OY Radiolinja AB
France	France Telecom, SFR
Germany	Deutsche Bundespost Telekom, Mannesmann Mobilfunk
Ireland	Telecom Ireland
Italy	SIP Italy
Luxembourg	P&T Luxembourg
Netherlands	PTT Telecom
Norway	Norwegian Telecom, NetCom GSM A/S
Portugal	Telecomunicacoes Moveis Nacionais (TMN), TELECEL
Spain	Telefonica Spain
Sweden	Swedish Telecom, Comvik GSM AB, AB Nordic Tel
Switzerland	Swiss PTT Telecom
Turkey	PTT Turkey
U.K.	Cellnet, Vodafone

Network (MCN) service for this purpose and plan to introduce it during 1993. It is expected that their GSM network will cover 90% of the population of Great Britain before the end of 1993.

Of the equipment manufacturers, Ericsson Radio Systems is well placed with orders to date. Motorola, Nokia, and Alcatel are supplying base station and switching equipment. Type-approved mobiles and transportables are available from several sources, and hand-portables are beginning to emerge.

6.7 THE FUTURE

GSM has created much interest worldwide, and it is anticipated that it will be taken up by many more countries outside Europe. Hong Kong has indicated that it will license competitive service, and the Far East in general is very interested in the standard. In total, there are now 23 countries in Europe and a further 20 outside Europe who have agreed to use GSM. The forecast is that there will be about 13 million digital cellular subscribers worldwide by 1996; over half of these will be using GSM equipment. This would be on target to meet the 10 million subscriber base by the end of the decade, which was the forecast in 1987.

Half-Rate Codec

As in the case of the full-rate codec, the selection of a half-rate codec is taking place by means of a formal competition between different approaches. The first evaluation, early in 1992, revealed that none of the candidates could meet the target performance specification. Since then, significant improvements have been made and there are currently two candidates that meet the requirements. The decision must now be made on the basis of the most suitable tradeoff between complexity and performance. The timetable is set to finalize the decision by November 1993 and complete the publication of the specification by the middle of 1994.

Frequency Band Extension

It is already being suggested that the 50 MHz allocated to GSM is insufficient to meet the anticipated demand. This is particularly true where the national authorities decide to license more than two competing operators. Of course, the opening up of the 1.8-GHz band to DCS 1800 could well alleviate this problem if the economics of running a network at that frequency prove satisfactory. Nevertheless, there is some indication that an eventual migration into the ETACS band is likely. This, it is argued, could be undertaken at the same time as the repositioning, or removal, of the few megahertz between the GSM and ETACS bands currently assigned to DSRR. Clearly, a contiguous band is more efficient than two separated bands. The use of ETACS is, however, likely to continue well into the next century and, in the United Kingdom at least, the extended version of GSM will not happen for many years.

Migration of Ideas From DCS 1800 to GSM

The PCN operators in the United Kingdom and now in Germany are expected to introduce new ideas and techniques to give them some competitive edge over their GSM rivals. Given the equivalence of the specifications, however, it is anticipated that practically anything done at 1.8 GHz can be migrated to 900 MHz. Thus, we can expect to see some reverse transfer of techniques as the systems battle for market share.

Reduction in Mobile/Hand-Portable Prices

Although mobile and hand-portable prices are currently higher than their equivalents for the TACS and NMT systems, we are already seeing competition causing the initial prices to fall. This is coupled with a number of manufacturers confounding predictions and achieving physical sizes that are comparable with the analog hand-portables. As further silicon integration is achieved and the market matures, we can expect to see hand-portables becoming no more expensive than today's analog models.

6.8 CONCLUSION

The GSM Committee adopted a pragmatic attitude towards its task. It is clearly a partnership between operators, regulators, and industry that is succeeding despite, or perhaps indeed because of, deregulation and the competitive pressures being introduced both in the PTTs and their local industrial base. We have seen drive, commitment, and, where necessary, compromise in the work thus far. If this is a model for future European initiatives, the continent can look forward with confidence to a technologically united Europe successfully competing with Japan and the United States in the global marketplace.

REFERENCES

[1] Herr, T. J., and T. J. Pleyrak, "ISDN: The Opportunity Begins," *IEEE Communications Magazine*, Vol. 24, No. 11, 1986, pp. 6–10.

[2] ETSI/GSM Recommendations, European Telecommunications Standards Institute, BP 152-F-06561, Valbonne, Cedex, France.

[3] Wakid, S., et al., "Coming to OSI: Network Resource Management and Global Reachability," *Data Communications*, December 1987.

[4] Natvig, J. E., "Speech Coding in the Pan-European Digital Mobile Radio System," *Speech Communication*, Special Issue, No. 1, 1988.

[5] Murota, K., and K. Hirade, "GMSK Modulation for Digital Mobile Radio Telephony," *IEEE Trans.*, Vol. COM-29, No. 7, 1981, pp. 1044–1050.

Chapter 7
Radio Equipment for GSM

C. Watson
Orbitel Mobile Communications, Ltd.

It has been said that the GSM system is the most complex radio system yet devised, and its specification is very demanding for the radio designer. This chapter explores the impact of the recommendations on the design and implementation of the GSM mobile stations, hand-portable units, and base stations.

7.1 INTRODUCTION

Chapter 6 discusses the approach taken by the GSM specification team and outlined the major features of the system. Spectral efficiency of the resulting system was one of the overriding design parameters due to the potential size of the user market. This requirement alone led to some very stringent design targets being placed on the radio subsystems. In this chapter, the challenges, constraints, and opportunities that faced the radio equipment designers are discussed in some depth.

Network operators and manufacturers were both represented in the GSM meetings from an early stage, and their respective objectives often led to requirements that were in conflict. Much discussion and compromise took place throughout, and it is believed that the resulting specification offers a good balance between the cost, the ease of implementation, and the performance levels attained.

Some of the design targets that have caused special concern are:

- Modulation Accuracy—To ensure efficient spectral usage and prevent receiver performance from being degraded by poor transmission sources, tight constraints were placed on frequency accuracy (±90 Hz) and phase trajectory (5 deg RMS).
- Multipath Equalization—Critical to the system is the performance of the equipment under Doppler shift and the time dispersion of the bursts. The equalizer design is

specified to cope with a Doppler shift equivalent to 250 km/h and a time dispersion of up to 16 μs.

* Power Control—Stringent requirements on power consumption in the terminal equipment and the need to minimize cochannel interference necessitates the use of power control of the transmitters. The lowest acceptable output power required to achieve a link is used. This requires the radio equipment to switch output power rapidly without causing switching transients.

In the following, the term MS will be used as an indication of all mobile station products, including the HPU, unless otherwise stated. The chapter initially discusses the issues that are common to both BTS and MS, and then examines the specific features of each in turn.

7.2 TRANSMITTER

In addition to the modulation accuracy and power control issue mentioned above, there are other factors that are important to the transmitter designer.

The choice of MS and BTS power classes can have a big influence on the network operators, since they will significantly affect the amount of infrastructure required to achieve the requisite coverage. Due to the requirement that the network operators be able to finely tune the radio coverage, the tolerance on the transmitted power for the BTS was made very tight. This meant that the accuracy required in the power control circuitry was very high. Too much tolerance in the BTS output power would either result in low output BTSs producing "holes" in the coverage or the system would be over-designed at extra cost.

A major requirement of the GSM system was that it should coexist with the current analog systems, NMT and TACS, described in Chapters 3 and 4. To enable this coexistence, the design must limit the amount of interference with other GSM and analog equipment. Specific design constraints for such emissions were imposed on the GSM system. The specifications of certain design parameters were then relaxed or tightened according to the power class of the BTS or MS.

The specific design constraints are detailed in the relevant GSM recommendations, but some of the more important factors applicable to all GSM radio equipment have been selected here for further consideration.

7.2.1 Modulation Accuracy

The modulation method selected for GSM is GMSK with a bandwidth-to-bit-period product (B·T) of 0.3. The choice of GMSK as the modulation method was based on a compromise of complexity and spectral efficiency. Quadrature phase-shift keying (QPSK) would give more efficient spectral usage, but would have limited the designer to Class A amplifiers in the transmitter path. These are very inefficient in terms of power consumption and power conversion of the wanted signal. GMSK modulation was selected because it provided a constant transmission amplitude output during the burst, and, hence, more efficient Class C amplifiers may be designed into the transmitter circuitry of both BTS and MS. The choice of modulation technique is more significant in the MS because of the battery efficiency required in order to generate an adequate standby time.

As GMSK gives a constant amplitude-modulated signal during the transmission burst, GSM specifies an allowed frequency offset and phase trajectory limit for both the MS and BTS. This is in terms of how much the actual transmitted burst is allowed to deviate from the predicted constant amplitude burst signal. The measurements of frequency accuracy and phase trajectory indicate the accuracy of conversion in the modulation process, and therefore place design constraints on the choice of modulation conversion. The specification is to ensure a correlation in performance of the MS and BTS across Europe. Link errors from MS to BTS or vice versa will be caused if the GSM specifications are not met by the transmitter outputs of either equipment. The receiver performance of both MS and BTS is specified under conditions of Doppler shift, multipath propagation, and interference. Operation must not be degraded beyond the limit specified in the GSM recommendations.

In addition to the limits of phase trajectory and frequency offset, there are requirements placed on the output RF spectral spread due to the modulation process, according to GSM Rec. 05.05. This specifies the maximum amount of RF energy allowed in other channels in order to not interfere with mobiles in the same sector. These modulation sidebands are particularly critical for the adjacent channels.

Measurements of modulation sidebands can be quite a complicated procedure due to the presence of switching transients. Under normal laboratory conditions, this measurement is not easily performed with standard test equipment, since the influence of the switching transients cannot be eliminated. This stimulated the test equipment manufacturers to produce equipment specifically for the measurement of the GSM transmitter parameters. However, as with all equipment for GSM, these tools were not readily available until most of the design had been finalized.

Figure 7.1 indicates the form of the GSM GMSK signal in terms of switching transients and modulation sidebands. This display is that which would be seen on a spectrum analyzer set on fast zero span with the measurement filter bandwidth set to 30 kHz. From the trace and the indication of its constituents, the problems of using conventional equipment to perform transmitter measurements can be readily appreciated.

7.2.2 Power Control

The GSM network is designed so that the MS is instructed to use only the minimum power level necessary to achieve effective communication with the BTS.

GSM defines eight power classes for the BTS transmitter to cover all five classes of mobile (0.8W to 20W), as shown in Table 7.1. For the BTS, the power output may

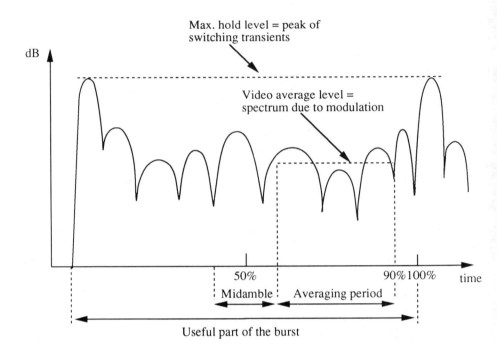

Figure 7.1 Time waveform of a GSM signal.

Table 7.1
GSM Transmitter Power Classes

Power Class	BTS Power (watts)	MS Power (watts)	MS Configuration
1	320	20	Mobile
2	160	8	Mobile/transportable
3	80	5	Hand-portable
4	40	2	Hand-portable
5	20	0.8	Hand-portable (MCN)
6	10		
7	5		
8	2.5		

be controlled, nominally in 2-dB steps, in order to provide better cochannel interference performance, which will allow a better service or a greater frequency reuse. The setting of power step 0 for the BTS will correspond to the relevant power class of the BTS transmitter. Both MS and BTS power control will be performed in 2-dB steps down from the level of the power class to a minimum of +13 dBm. The power output level of the

MS, however, must be controlled in a monotonic sequence of fifteen 2-dB steps on command through the SACCH from the BTS. It is also important to note that during a link to the BTS, the MS can only be instructed to change the power level by one step between subsequent time slots.

The use of minimum transmitter power in accessing the network also helps to increase the battery lifetime of the MS and reduce interference. Battery life is of paramount importance when considering HPU configurations, since the battery capacity is limited by its size.

An unwanted side effect of power control is the out-of-band radiation caused by switching transients. These predominate in the RF spectrum above an offset of 400 kHz away from the carrier as they become more dominant above the modulation sidebands. Maximum limits of spurii due to the wanted carrier are specified in the adjacent channel and farther out. A very stringent limit on switching transients is specified in the associated receive band of all the GSM equipment for the obvious reasons of self-deafening. For the MS, the limits set for switching transients are to minimize the interference with cosited MSs being used in the same sector.

In the receive band, the specification limit for switching transients is an absolute limit of -103 dBm for the BTS in the frequency band of 890 to 915 MHz, and -76 dBm for a Class 1 MS and -84 dBm for all other MS classes in the frequency band 935 to 960 MHz.

By carefully controlling the ramp-on of the transmitter as well as the power level the spectral interference with other GSM equipment can be minimized.

Figure 7.2 illustrates the required level of control in terms of the timing of the transmitter burst, since all GSM equipment must ensure that its performance fits within this "mask." New methods of power control have been developed in order to maintain stability within the control loop over the required dynamic range. Any instability will result in the breaking of the time mask and hence in the failing of the spectral requirements on switching transients. It may not be necessary for the BTS to power up from zero, since the previous time slot may be occupied. The BTS will only be required to change the relative level of its output in the new time slot. The MS, however, must power up from the "off" condition, that is, zero output power.

7.2.3 Spurious Emissions

Spurious emissions refer to those emissions outside the spectral regions specified by modulation or switching transients. GSM specifies maximum limits for these over the range of 9 kHz to 12.75 GHz. This specification is designed to minimize interference with other communication equipment caused by unwanted transmissions such as modulation products, clock harmonics, and the like. In order to measure spurious emissions, it is necessary to use a five-pole synchronously tuned measurement filter. This is to achieve a repeatable impulse response on the measuring equipment.

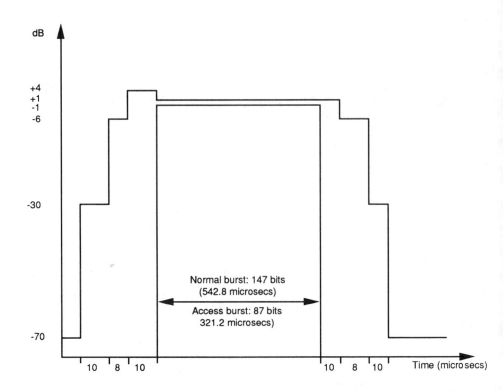

Figure 7.2 Burst timing mask.

The out-of-band spectral limit is specified to be an absolute level of −36 dBm up to 1 GHz, except for the stringent requirements of the receive band mentioned above. There has been a lot of concern as to whether this absolute limit is too high. Interference observed with video terminals, telephones, and hearing aids is due to the pulsed carrier transmission, however, and not due to spurii.

At frequency offsets from the carrier of more than 1.8 MHz in the range 9 kHz to 1 GHz, the resulting energy must not be more than an absolute level of −36 dBm. This requirement can be as much as 79 dB down from the carrier peak power of a Class 1 MS or 81 dB for a Class 1 BTS. This level is, however, relaxed above 1 GHz to −30 dBm for both MS and BTS. This results in tight specifications for the filtering of the transmitted pulse, and consequently costly duplexers are used rather than transmit/receive switches in MS products. For the BTS, highly tuned cavity resonators are required for filtering which are both bulky and expensive.

Note that although the spurious emissions have been given for the mobile and BTS in transmit mode, the limits apply equally to the GSM equipment in receive mode; but in this case they are more stringent, being −57 dBm for the frequency range 9 kHz to

GHz and −47 dBm for 1 GHz to 12.75 GHz for all equipment, because of cositing MS
nd BTS equipment.

With the equipment in idle mode, no interference is experienced by other equipment,
rovided the GSM specification is adhered to. Interference only arises once the MS
ansmits to the BTS.

.3 RECEIVER

he receivers for both the BTS and MS, but not the HPU, have a minimum required
ynamic range performance of 94 dB. This allows the MS and BTS receivers to work
ver the range of −10 to −104 dBm, relaxing to −102 dBm for HPU. The measurement
chnique for the receiver performance assumes that there are no errors in the transmitter
ath. Then a loopback facility is invoked whereby the information transmitted back is
e same as that received. From this the error rate of the link is calculated by measuring:

- Bit error rate (BER), where a calculation is made of the bit wrongly received
  compared to the data bits sent.
- Frame erasure rate (FER), where a measurement of the result of the 3-bit cyclic
  redundancy check for speech channels is made along with the bad frame indicator
  (BFI). For signaling channels, the FER is a measurement of the errors in the block
  code used.
- Residual BER, where a measurement is made of the frames which have not been
  declared erased.

hese measurements are specified in order to maintain the performance of the GSM
ystem under extreme conditions and in the presence of strong carriers, wanted or other-
ise, modulated or not. A number of other parameters, such as blocking, intermodulation,
nd cochannel rejection, are also specified in GSM Rec. 05.05. These limit the design
ptions available for the allocation of gain and filtering in the receive path.

.3.1 Equalization

ne of the major limiting factors of the GSM equipment is in the performance of the
qualizer. The equalizer is a feature of GSM and improves the overall performance of
e system.

Due to the fact that the modulation rate is comparable to the radio channel bandwidth,
vere intersymbol interference (ISI) and echo due to multipath propagation may result.
he receiver has a certain amount of rejection of unwanted signals, provided the frequencies
e sufficiently far away (one channel). The presence of multipath necessitates an equalizer
 order to overcome the problem of ISI and echo.

Channel performance must be maintained in the presence of multipath, with possible
ath delays of up to 16 μs and frequency shifts corresponding to speeds of 250 km/h
ased on the high speed of the French *train a grande vitesse* (TGV)).

Equalization is performed by synchronizing with a fixed training sequence consisting of 26 bits, which arrives in the middle of the burst. A correlation is then made of the known sequence against that received. From this correlation, a better prediction can be made of the correction process required for processing the remaining data than if the training sequence appeared at the beginning or at the end of the burst. This method reduces the amount of error during the pre- or postprocessing of the data. Most equipment manufacturers have based the correction process on the Viterbi equalizer, although the direct feedback equalizer (DFE) has also been suggested. Synchronization itself will compensate for up to 233-µs absolute delay.

The equalizer must also compensate for the Doppler shifting of the burst's frequency due to the movement of the MS. For the BTS, this measurement is performed by the equalizer in order to determine the actual position of the MS transmit window in relation to its distance from the BTS. This is so that the BTS can advise the MS to change its timing advance in order to synchronize to the correct time slot allocated by the BTS. MS timing relative to the specified time slot boundaries may then be adjusted over a range of up to 63-bit periods in steps of 1-bit periods (3.7 µs). Therefore, the MS transmit pulse can be adjusted to be kept inside the allotted time slot at the BTS. The guard space around the time slot is 30 µs, which when added to the timing adjustment allowed, constitutes a time delay equivalent to a one-way path length of 35 km.

This device is very complex and the nature of the operation and performance of the MS depends heavily on the implementation and specification of the equalizer. Solutions have been produced using digital signal processing (DSP) devices, but size and power consumption constraints within the MS have generally led to a custom silicon approach.

7.4 TRANSCEIVER PERFORMANCE

This section discusses those parameters that are not specific to the receiver or to the transmitter of the GSM equipment.

7.4.1 Frequency Agility

Chapter 6 explains why the GSM transceiver must be frequency-agile. The mobile, during one frame of 4.6 ms, must be capable of performing one transmit, one receive, and one monitor operation on different channel frequencies at what could be opposite ends of the GSM frequency band. This places stringent demands on the tuning time and accuracy of the synthesizer. Given that the MS has to frequency hop, it was decided to take advantage of this and specify hopping as a method of randomizing cochannel interference. This means that the BTS must also hop, and in its case the frequency can change every time slot. This gives a hopping rate of 217 hops/s. The different requirements on the MS and BTS can lead to different solutions, and the BTS options are discussed in more detail in Section 7.7.2.5.

There are two methods of hopping used by the GSM system, which apply to two different general situations.

7.4.1.1 Monitor of Surrounding Cells

The mobile is required to monitor surrounding cells during its normal operation, and as mentioned above, it also has to hop between the three different windows (receive, transmit, and monitor). This method of hopping was the only mode of operation adopted in the original system setup because of hardware and software limitations. As functionality increased, the hopping method explained in the next section was invoked.

During idle and traffic modes, the mobile continually measures the signal strength and quality of the link to the neighboring cells by means of bit-error-rate measurement. This information is recorded in a table in the mobile's memory, and is then reported back to the serving cell where it is used to determine the movement of the mobile within the cells and hence the requirement for handover or cell reselection.

The aim of this procedure, in traffic mode only, is to enable the base station controller (BSC) to make a decision on whether to hand over the mobile to the next cell as the signal level and quality reduces. The decision to perform a handover is taken by the network in order to ensure that there is spare capacity on the receiving BTS. There may be cases when capacity is not available on the neighboring cell, so the mobile reestablishes itself on the old cell and another attempt is initiated by the network for handover to another cell.

In idle mode, the recorded cell measurement reports of the neighboring BTSs are used by the MS to initiate a cell reselection to a stronger channel of its home network, if appropriate. A location update by the MS will only occur if the MS changes MSC or BSC cell boundaries.

7.4.1.2 Frequency Hopping

The second method of frequency hopping is more commonly associated with military radios. It is frequency hopping in order to maintain good communications in areas of slow fading and to maintain the security aspects of the system. Without hopping, the mobile's performance will be seriously degraded in areas of deep fades.

Fades occur when there is a loss in signal reception due to geographical features such as valleys or hills, or due to objects such as buildings or even large metal objects such as aircraft, interfering with the signal path, causing the original wanted signal to be attenuated or canceled out.

As with all mobile radio, the MS will pass through areas of fade and poor reception, and by invoking frequency hopping, there is less chance of losing the radio link in these areas. This is especially important when trying to perform a handover to another cell. Decisions to hand over are made in areas of poor reception, and a lot of signaling is required in order to perform the process. If the fade is a deep one, then the communication link required to signal the handover may not be achieved, and the link will be lost. Frequency hopping allows the opportunity to maintain the link by moving onto another frequency before the link is completely lost.

Due to the performance of the equalizer and channel coder, the error correction process can manage with a loss of information of up to one frame in five, as will occur with deep fades. By the time the fade affects the radio link, the mobile will have hopped onto another frequency and reestablished the signaling.

The rate of hopping used in the GSM system is 217 hops/s. On registration, the MS is instructed on the hopping algorithm to be used for the duration of the call. This algorithm may be changed at any time during the progress of the call, especially when the MS performs handover. It may be that the mobile will hand over between hopping and nonhopping channels, depending on the network arrangement; hence the requirement for the MS to support hopping.

7.4.2 Channel Decoding and Encoding

There are many types of channels specified within GSM, such as speech, control, and data channels. Each channel is allocated a convolutional code according to Rec. 05.03. Convolutional coding was chosen for GSM because it is superior to binary block coding when dealing with random errors caused by the radio link path.

The reason for supporting different dedicated channel types is that specific control channels are always available for the signaling and link management information required to maintain the GSM radio link, even when the network is congested with speech.

The coded signaling or speech information is then interleaved. There is sufficient redundancy within the interleaving process of the GSM signal structures to allow for one frame in five to be lost without a significant loss in quality. For the decoding process, a Viterbi algorithm was considered the best solution for GSM.

As with the equalizer, because of the large amount of processing required, DSPs as well as custom devices will provide a solution. The choice will depend on the space available and the power consumption. It is possible, depending on the architecture chosen, that the equalizer and channel coder may be combined as their operation is interactive in the receiver path. The channel coder's size and complexity are similar to the equalizer's.

7.4.3 Encryption

A major problem with the analog cellular radio systems has been the ease with which telephone calls can be monitored using standard radio communication equipment. To overcome this problem, the use of digital encryption has been specified by GSM Rec. 03.20. The use of encryption algorithms within the channel coding, along with the SIM card, ensures that security is high. A SIM or Smart card is a personalized identity card that allows the user to insert the card into any GSM phone and make calls. The calls will then be charged to the SIM card holder. Not only is the GSM speech encrypted, but the signaling is too, so any person trying to eavesdrop will therefore have to decipher the signaling information. This requires that the listener trace the progress of the whole call in order to find out the hopping sequence and cipher key set.

Whereas the use of ciphering ensures the security of the radio link, the SIM card provides security of the MS equipment in the event of theft. Both the SIM card and the MS equipment have unique identity numbers that are transmitted to the network. This information helps to police the network for unauthorized users.

7.4.4 Speech Coding

The speech coder selected for GSM is a linear predictive coder with long-term prediction which produces a 13-Kbps data rate, as shown in Figure 7.3. The coder works with 20-ms blocks of speech samples consisting of 260 bits, which leads to a protected full-rate speech data rate of 22.8 Kbps. This then produces a subsequent modulation data rate of 270.833 kHz. The coding and encoding algorithms are specified in GSM Rec. 06.10.

For the BTS, remote speech transcoding may be performed. In this case, the speech coder is situated at the MSC or BSC instead of at the BTS. The interface will then be straight into the PSTN. The advantages and disadvantages of this method will be discussed later in Section 7.7.2, where the BTS is given fuller consideration.

In a mobile, speech data at a rate of 13 Kbps is block-diagonally interleaved (see Figure 7.4). This allows recovery from a fast fading signal and selective loss of a transmission burst when frequency hopping is employed.

Selecting a suitable device for the speech coder will depend on how many features are to be included in the one device. If the analog audio functions are to be incorporated, a mixed CMOS-bipolar technology would be required. However, standard speech processor devices are now available, and if these are used, the speech coder can be conventional CMOS. Such an approach increases the device count, however.

In the future, there will also be the requirement to support half-rate coding for data and speech transmission up to 4.8 Kbps. This will allow for an approximate doubling of the current maximum capacity of users in the GSM system. The technology for half-rate coders is not yet available, but then the system is hardly congested at present.

7.4.5 Short-Message Service and Data Services

Further features of GSM system include the use of SMS and the support of data services. For these facilities, the channel coder must also be capable of supporting data rates of up to 9.6 Kbps, along with its current signaling capabilities.

SMS allows the user to receive messages in much the same way that a pager operates, provided that the user has subscribed to this particular supplementary service.

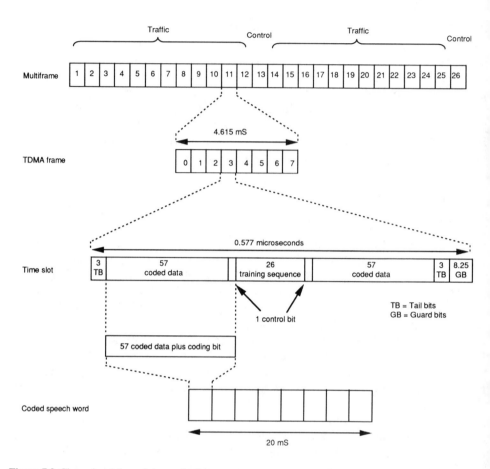

Figure 7.3 Channel and time slot organization.

The MS can also initiate, as well as receive, messages from other users. These messages are stored in the SIM card for retrieval later when they can be displayed on the standard MS display. This feature adds a further requirement to the MS display functionality, which it would otherwise not have to support: the presentation of full ASCII alphanumeric messages.

7.5 SIGNALING AND SOFTWARE REQUIREMENTS

The construction of the signaling channels for GSM is complex because it contains all the features and control protocols necessary for an ISDN system. The main reason for defining specific dedicated channels in GSM is so that channels will always be available for link management and control requirements. This ensures that channel capacity will

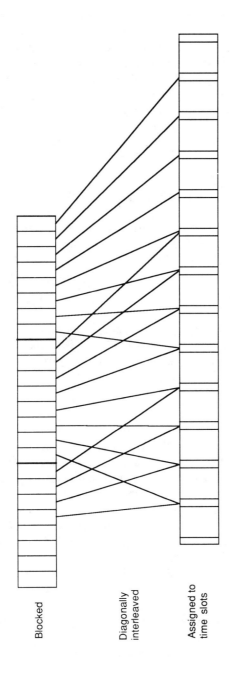

Figure 7.4 Block diagonal interleaving.

always be available to carry important information, especially in cases of fast handovers. Different encoding algorithms are also applied to different channels, making it difficult for eavesdroppers to monitor the progress on the channel link. Chapter 6 specifies in more detail the signaling and data structure of the GSM network.

The signaling requirements of the BTS are the same as those for the MS, but, in addition, the BTS requires far more software control in order to maintain large databases for the MS subscriber data and information, as well as cope with fault detection within the system. The BTS software is designed to improve remote service capabilities and repair in order to maintain a satisfactory network service. Network management software is discussed in Chapter 6.

7.6 MOBILE AND HAND-PORTABLE STATIONS

A number of technical and marketing constraints were placed on the MS even before the first design ideas hit the backs of envelopes, and some of these issues are discussed in the following section.

7.6.1 Market Constraints

The MS products have to compete against existing networks and their products if people are going to be encouraged to move onto the GSM system. The market thus determined aspects such as size, cost, and battery lifetime with reference to the current performance of analog systems.

7.6.1.1 Size and Cost

The resulting GSM MS, one of which can be seen in the photograph in Figure 7.5, will be compared to existing analog products. The TACS and NMT markets are into second- and third-generation products, and these are considerably smaller than the first-generation GSM products, especially with regard to hand-portables.

The average MS user is not aware of the complexity required of the new system, and they should not need to be. Therefore, all the users can compare is the size of the MS and the features provided for the GSM system against the current analog networks, TACS and NMT. Cost is also of major importance.

In order to meet these size and cost targets, the use of custom-designed silicon devices is widespread, with an ever increasing requirement for greater silicon integration despite the large investment required.

Advances in chip technology will also help to bring the cost of future generation equipment down. A 0.8-μm CMOS is currently available for silicon design, and the common current design rule in MOS devices is that the device density will double every two years.

Figure 7.5 A GSM mobile station.

To implement GSM within the required size and cost constraints, two design and implementation options are available: a hardware solution and a software solution.

Due to the complexity of the GSM system, the use of standard devices was not possible. Therefore, a lot of investment in new chip designs would be required if a hardware solution was chosen. The risk was that the GSM recommendations were not stable in the early design phase, so the design may have had to be written off if a key design feature was changed by GSM. This risk was minimized by the presence of MS manufacturers on the GSM review committees. Two or more iterations of silicon might have been required to ensure complete functionality according to GSM if specifications were changed significantly or the implementation was not fully functional.

A software solution to GSM involving a number of DSP devices would have provided greater flexibility in the event of design changes, but would have required much greater software control, resulting in very complex software. The power consumption of such a solution is likely to be higher than for a hardware design, but for early prototypes, this was not really a consideration. Such MS terminals were available for testing much later than a hardware solution was because of the amount of software necessary to achieve even a basic functionality, but when DSP prototypes were available, their functionality could be rapidly increased.

The merits of both methods are about equal, taking everything into account, and the final choice depends on the capability available within the manufacturer's design team, along with the company's specific objectives for GSM. The provision of early prototypes for system evaluation may also have influenced the design methodology. Early prototypes, based on hardware solutions, helped to validate specifications and the development of specific GSM test equipment.

7.6.1.2 SIM Cards

GSM adopted the concept of the SIM card, the details of which are given in Chapter 6. This facility has already been available in Germany in their analog system. This Smart Card provides additional security for the user, with the requirement of a PIN and a subsequent password. However, it also introduced a design constraint for MS manufacturers. Two versions were specified, a standard credit card size and a cut-down version for semipermanent installation in the MS. The location of the SIM card reader, plus the choice of whether to support the large or small card, has significant impact on the size of the MS as well as on the design of the vehicle mounting kit for the transceiver. Should the card be mounted in the handset or the transceiver case, and where should it be located for easy access?

7.6.1.3 Manufacturing

The manufacturer's capability and accuracy in the placing of components during production has a great impact on the cost and size of the product, along with the amount of automatic and manual setup and testing that is required to be performed on the product before it can be delivered to customers. The GSM HPU has forced manufacturers to achieve improved packing densities and production test methodologies.

GSM should, however, provide the mass market required to bring down the actual cost of the MS production, while recouping the enormous investment cost.

7.6.1.4 Power Consumption

Power consumption is of major importance, especially when similar products on the market are being compared. The final choice of battery technology and capacity is a compromise based on the size and weight of the MS and the perceived customer requirement for battery life. The general aim for all terminal designers is to achieve a typical 8-hour-day use from the MS as a minimum.

The designers of the MS are thus limited to the use of low-power CMOS devices. As silicon technology progresses, the introduction of lower-supply-voltage devices will bring significant improvement in the power consumption figures. However, the performance of the RF circuits cannot be compromised, and it remains to be seen whether a cost-effective solution can be found.

7.6.2 Technical Considerations

Apart from the marketing constraints, GSM required new hardware and design techniques in order to achieve the specifications, as indicated above. Figure 7.6 shows a block diagram of the basic MS structure.

7.6.2.1 Frequency Stability

The presence of an equalizer in both the MS and the BTS helps to provide a good fixed reference for the radio communications under all propagation and interference conditions.

The required frequency stability of the BTS must be better than 0.05 ppm in order for frequency correction by the MS to be performed against the BTS transmissions. This allows a lower-cost alternative crystal reference (typically 3 ppm) to be used with GSM MS products compared to the high-specification references (typically 1 ppm) used in current analog terminals, because the BTS adjusts the MS reference in accordance with its high-precision reference.

The tuning range of the crystal required will determine the speed at which the MS will be able to access the network. Too large a range will mean that the MS takes longer to search for a network (i.e., to find a BCCH frequency). Therefore, the crystal accuracy is traded off against the size of the crystal and the speed of searching all channels for BCCH frequencies. It is unacceptable for an MS to take more than a few seconds to access its home network.

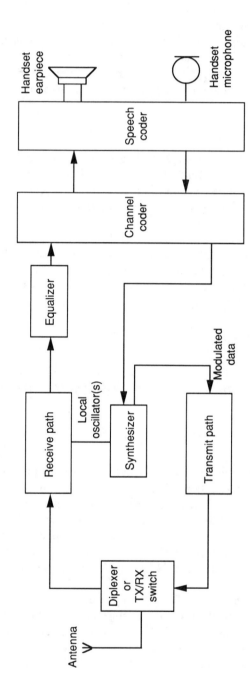

Figure 7.6 MS architecture.

7.6.2.2 Synthesizer and Modulation Techniques

Another technical constraint is in the choice of modulation technique.

First, in order to solve the problem of tuning over the two GSM frequency bands, it is common to use two synthesizers. However, for the MS, size is of great importance, not to mention the component cost of such a decision and the potential repercussions with problems of spurii due to multiples of both voltage-controlled oscillators in each synthesizer interacting with the 13-MHz system clock. Filtering out these modulation products would also become very difficult in order to achieve the specification on spurious emissions.

A more refined and cost-effective approach is to use a single synthesizer with a fast phase-locked loop to cover both frequency bands. This means that the synthesizer must be capable of settling within a few hundred microseconds. Using the same synthesizer for the local oscillator both in the receive and transmit paths, the design must provide sufficient bandwidth to accommodate the modulation required in the transmit path, but be narrow enough to be a clean local oscillator for the receive path. The two requirements on bandwidth may limit the synthesizer settling time and design.

To meet these requirements requires a good stable synthesizer with little or no residual amplitude modulation (AM) on the burst, such that after amplification, with what is likely to be a Class C power amplifier, no resulting phase AM results. GMSK is a constant AM, and as such, problems should only arise in the implementation of the modulation process. There are a number of possible modulation techniques available, three of which are discussed below.

Digital Interpolation (Digiphase) Modulation. This is a very accurate method of modulation. With this method, the required modulated signal is accurately represented using a number of accumulators around which a fast-settling phase-locked loop operates.

IQ Modulation. With this method, a quadrature (sine and cosine) representation of the data stream is produced and then upconverted to produce the modulated signal. Many problems may arise if the two channels are not perfectly matched, and the modulation accuracy can be degraded. In this instance, analog processing must occur before the modulation process takes place, which may result in uncertainties. A digital version of this method, however, reduces the uncertainties and results in improved modulation and upconversion.

Frequency Modulation. FM is the most commonly used method for generating carriers in mobile radio systems. When used for a GMSK modulated signal, FM would require a high-bandwidth, high-resolution synthesizer in order to achieve the required modulation at the different frequencies. This would have its own problems in terms of accuracy, bandwidth, and stability, requiring great care in the design stages.

The final choice of modulator for the MS plays a large part in meeting the modulation accuracy specification and the resulting phase and frequency parameters as required by GSM. This is the result of continuing development.

As discussed earlier, there is a GSM specification for the spurious emissions allowed in the frequency spectrum 9 kHz to 12.75 GHz. If using half- or quarter-rate frequency synthesizers, care must be taken over the presence of spurii at the original frequency after final power amplification. This is especially important when GSM has to coexist with NMT systems that operate at 450 MHz. The out-of-band spectral limit is an absolute level of −36 dBm up to 1 GHz. Filtering at the antenna socket can help to minimize the levels of spurious emissions, as well as good design and selectivity of the frequency response of the power amplifier. The use of a diplexer at the antenna socket of Class 1 and 2 MS is favored, especially for reducing the spurii in the receive band of the mobile, where the specification is much more stringent. For the HPU, the output power is much lower, and a transmit/receive switch may be sufficient as the antenna combiner.

7.6.2.3 Transmitter Control

Careful consideration must be given to the method of power control of the power amplifier (PA) and to the class of PA used. A Class C amplifier may turn any residual AM on the modulated signal from the synthesizer into phase modulation on the transmitted burst. This causes the resulting burst of data to be outside the GSM specification. This "failure" can be in terms of modulation accuracy (i.e., phase and frequency error) or in the resulting modulation or switching sidebands.

The shaping requirements placed by GSM on the burst ramp-up of the power amplifier are that it must be controllable over a possible range of 30 dB. The PA must also be capable of being switched to a stable state at +7 dBm. This necessitated many consultations with power amplifier manufacturers, for the purpose of producing specific modules for the GSM system. They need to be power-efficient, cheap, and capable of being turned on and off quickly. The choice and method of power control systems also imposed slightly different requirements on PA manufacturers, and as a result, modules specific to GSM were slow to develop.

A method of power control is shown in Figure 7.7. This method allows the shaping to be applied to the random-access and traffic transmit bursts. It also allowed for flexibility and fine tuning to be performed to compensate for changes in the control loop around the transmitter as the PA modules were being developed.

Unlike the BTS, the MS must always power up from the "off" position, and, hence, the residual output power level must not be more than −70 dBc or −36 dBm, whichever is higher.

Temperature stability around the transmitter's control loop can also be a problem as the PA warms up. The control loop around the power control circuitry must be capable of overcoming any temperature variations, along with any glitches caused in the method of pulse shaping chosen, without going unstable. The GSM-specified operating temperature range is −20° to +55°C.

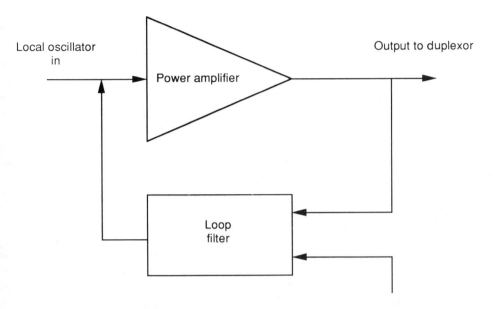

Figure 7.7 Power amplifier control.

7.6.2.4 Receiver IF Arrangement

The choice of the intermediate frequency or frequencies (IF) is based on the compromise of availability of standard products with the generation of multiple interferers caused by harmonics of the 13-MHz clock reference. By splitting up the overall gain and filter requirements amongst the IF stages, a simpler and stabler receiver operation is possible. Various calculations had to be performed so that the correct MS performance could be obtained with the chosen IFs. The final performance of the MS receiver is to achieve a noise figure of 10 dB at maximum receiver gain. The architecture used will vary greatly between MS designers and manufacturers, which can initially make components expensive. The final choice of IF will then dictate the conversion method required to achieve the 270.833-Kbps data rate required by the equalizer circuitry.

7.6.2.5 Filter Requirements

Sufficient filtering must be provided at each IF stage in order to achieve the requirements of adjacent channel rejection, out-of-band spurii rejection, and blocking. The specification of the channel and roofing filters must not be too tight, since this will make the components expensive.

Because of the different filter technologies, the choice of IF will have an impact on the cost of the MS. For example, the use of transmit/receive switches may be preferable to costly diplexers at the antenna socket if sufficient filtering is provided elsewhere in the design.

7.6.3 Software Control

Software control will predominate in the user's man-machine interface (MMI). Good software control will produce a user-friendly interface with good overall control and good power consumption efficiency. One effect of software control is the use of staggered paging groups to call the MS. This allows the use of ''sleep'' modes within the MS, which will help to extend the battery life, since the ''power-hungry'' features such as the liquid crystal display (LCD) backlighting will be turned off when the equipment is in standby mode.

DTX is also a feature of GSM, whereby speech is only transmitted when there is speech available to transmit. This helps to reduce RF interference in cosited MSs. Voice activity detection (VAD) is used to initiate this switching process. To help maintain a high level of intelligibility, *comfort noise* has been introduced at the receiver during the intervals that the speech has been cut. Comfort noise is a low-level background noise, based on the statistics of the acoustic noise at the transmitter, which reassures the listener that the radio link is still present. Under normal circumstances, a digital link would be completely silent. The use of DTX also helps to save power and generate a longer battery lifetime.

7.7 BASE STATION

7.7.1 Specification Constraints

Although some of the design features for the BTS have the same complexity as those for the MS, there are some specific parameters that place stringent requirements specifically on the BTS architecture and design.

7.7.1.1 Coexistence With Analog Systems

Currently in the United Kingdom, the 25-MHz GSM frequency band is shared between the TACS and GSM systems. TACS uses the lower 15 MHz of the frequency allocation, and GSM uses the top 10 MHz. ETACS is an extension band for TACS in the United Kingdom that uses the 16 MHz just below that allocated for GSM and TACS. Similarly in Scandinavia, NMT uses the 900-MHz band. GSM equipment must coexist with these analog systems with minimal interference.

Due to the presence of TACS and NMT signals, specification of the rejection of intermodulation products and spurious emissions is very tight within GSM to enable this coexistence. This is because of the fact that although the GSM BTS sensitivity level is −104 dBm, the corresponding sensitivity of the analog system is −123 dBm due to the narrower bandwidth.

By cositing GSM BTSs with TACS or NMT base stations, the intermodulation interference can be reduced; but additional problems arise due to the sharing of antennas, power supplies, and network transmissions.

One significant difference between analog and digital systems is that each GSM BTS must provide multiple traffic channels. The BTS must be capable of providing at least 8 full-rate channels per single piece of transceiver equipment. This will be increased to 16 as half-rate channels are introduced. Due to the nature of TDMA, the combination of carriers at the antenna is simpler. For example, to provide 16 full-rate GSM channels, only two GSM carriers need to be combined, since each supports 8 channels. For "single channel per carrier" analog systems (AMPS, TACS, and NMT), it would be necessary to combine 16 carriers, thus introducing additional complexity and inefficiency.

Another benefit of a TDMA system is that higher power transmitters may be used without the heating effects caused in continuous wave (CW) analog systems. Pulsed transmitters can thus support a longer mean time between failures (MTBF), provided suitable components are chosen to cope with the TDMA nature of switching the transmitter on and off.

There will always be a problem with competitive digital networks in the same area because the planned cell boundaries and frequency allocation will not be the same for both networks. As a result, cosite interference of MS will occur.

7.7.1.2 Frequency Accuracy

The BTS provides the system time and frequency reference in the communication between the MS and the BTS and must therefore be of high accuracy and stability.

Good timing accuracy is required in order for the MS to synchronize reliably with the BTS while moving about and to allow the MS to use sleep modes. The BTS will control the timing advance function of the MS, which is used to force the MS onto the correct time slot at the BTS receiver. If this feature were not imposed, each time slot would have to have sufficient guard space around it to allow for the possible maximum round-trip time delay between BTS and MS, resulting in the inefficient use of the spectrum.

A high degree of absolute frequency accuracy is required in the BTS, as discussed previously. The stability of the BTS reference is derived from a digital transmission system using a very long time constant in order to filter out short-term jitter. The digital transmission network itself is locked onto a rubidium standard.

The accuracy of the frequency reference within the system will determine the speed of handover by the MS. By providing a good time reference, the MS, when performing measurement reports on neighboring cells, will have a more accurate measurement of absolute timing required for new channels so that, during a handover condition, the MS will be able to synchronize more quickly with the next channel. This can be vital in areas of deep fading experienced by the MS.

7.7.1.3 Dynamic Range

As already mentioned, there is a potential problem in the "near-far" relationship between two mobiles and a BTS (Figure 7.8). This is where one MS is located effectively under the antenna and the other located at the cell boundary. With power control invoked by the BTS, the stronger MS signal will interfere with the weaker signal from the furthest MS. The BTS must be able to separate and decode the two signals, which may have a signal strength difference of up to 116 dB (i.e., +13 to −103 dBm) and may also be on adjacent time slots.

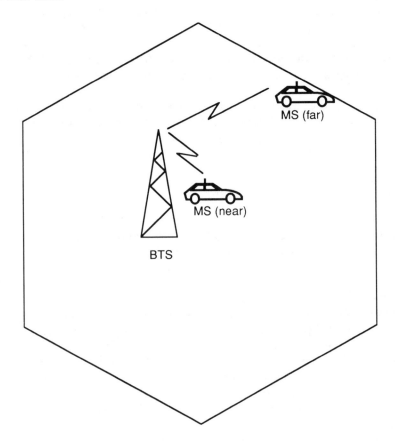

Figure 7.8 Near-far MS problem.

By using the power control capability of the MS and by careful use of assignment channels, the occurrence of this situation may be reduced. It may therefore be sufficient for a dynamic range of 80 dB to be provided (i.e., blocking (−23 dBm) minus sensitivity (−103 dBm)). Interference will also arise in the MS due to the presence of analog signals, but this issue has already been discussed in Section 7.2.2.

When the MS first establishes itself on the network, the automatic gain control (AGC) of the BTS must be very fast acting. It must be able to deal with a random-access burst (RACH) from the MS requesting a channel. The RACH is an unsynchronized channel burst that may come at any time, and the BTS must be capable of capturing and decoding it with 95% accuracy. As an added problem, the access burst power is unpredictable and may be at any of the 16 power levels from the MS, which equates to a 30-dB range. However, since the power level used by the MS is part of the cell broadcast information, some of the unpredictability is reduced. Handover to another traffic channel will also provide the AGC with work to do in order to stabilize the traffic link on a new channel in an error-free mode.

7.7.1.4 Maintenance of Service

The requirements of GSM are for a very high availability of service, which imposes particular requirements on the BTS equipment design. The system design must allow for remote (centralized) control over equipment and the gathering of fault reports. The proposed MTBF for all equipment modules is expected to exceed 2.5 years in continuous operation.

Provision of redundancy for all critical modules (e.g., spare transceiver, transmission, and controllers) allows for extra service availability. Under these circumstances, for example, the transmitter frequency or power class may be reconfigured if one stage of the amplification process fails, without losing the BTS functionality. Remote system support features include download of software to allow for ease of system upgrades, and configuration changes.

All communication systems need high reliability and serviceability, and GSM is no different. Sites may not be easily accessible due to geographical terrain, and redundancy and self-testing is of paramount importance, as are battery backup facilities.

In many cases, channel frequencies at the BTS may be remotely configurable, therefore permitting changes to be made to the frequency configuration of the system without visiting the BTS site. This can be a significant cost advantage over analog systems, and will increase the overall system operation flexibility.

7.7.2 Technical Considerations

7.7.2.1 Remote Transcoding

GSM allows transcoding of the speech data to be performed remotely from the BTS or network site. Speech transcoding is normally performed by the BTS after the channel code at the rate of 13 Kbps. The GSM network following the BTS requires the speech data to be pulse-code modulated at a rate of 64 Kbps (i.e., normal telecommunications standard).

Location of the transcoder at the BSC or MSC permits the multiplexing of four channels at 16 Kbps (allowing for overheads above 13 Kbps) into a 64-Kbps data stream, providing a saving of 4:1 in the transmission costs.

7.7.2.2 Signal Concentration

Each transceiver in the base transceiver station typically supports eight full-rate channels. The channels are used not only for speech, but also for the signaling and control, BCCH, FACCH, SACCH, and slow dedicated control channels (SDCCH), as required for the GSM radio link.

In a typical network, at maximum capacity, the BTS would require nine 64-Kbps transmission channels, eight for traffic and one for signaling. By performing remote transcoding, this requirement is reduced to only three channels, two for traffic and one for signaling. This implies a 50% signaling overhead. It is also possible, however, to multiplex the signaling traffic from three or four transceivers onto a single transmission carrier, thereby providing cost savings for the operator.

7.7.2.3 Power Output and Control

The BTS output power for GSM is specified to be measured at the input to the transmitter (TX) combiner. The choice of combiner and feeder loss to the antenna can have a significant effect on the resulting output power from the BTS, and hence in the design of the network coverage. Therefore, some manufacturers choose to specify the effective transmitted power based on assumptions of combiners used. The most typical choice of BTS power is a Class 3 or 4 BTS, giving a maximum output power of 40W or 80W at the input to the combiner. This results in approximately 10W to 20W at the antenna.

From the network coverage predictions, network operators are able to fix the maximum output powers from the individual channels in the BTS in order to minimize interference with other cells. The power control steps beyond this preset level, set at installation, can be used by the BTS during calls, with a maximum change of one step between time slots. Note that any preset power can alter the effective dynamic range available, since the minimum output power, allowing for losses in the combiners, must not be less than the minimum output power from an MS.

The actual benefit of this fine adjustment of power control during the progress of a call is subject to some debate, since the early BTSs were unable to perform adaptive power control. Changes in power level by the BTS will affect the amount of cochannel interference and, therefore, may lead to instabilities in the system. It is believed that the real benefit of dynamic transceiver control will be found when frequency hopping is introduced.

7.7.2.4 Combiner Technology

The BTS is made up of a number of transceiver units that have to be multiplexed together onto one antenna. (Planning permission, cost, and performance constraints make it highly undesirable to use more than one antenna per transceiver.) There are a number of methods for combining the signals, explained in the following.

Cavity Combiners

Cavity combiners use frequency-selective cavity filters in order to combine the transmitter signals. They provide a low-loss signal path for the transmitted signal that is also heavily filtered. This filtering can help to reduce problems caused by intermodulation products or wideband transmitter noise, as with the MS in the near-far situation in Figure 7.8. When combining multiple transmitted signals, the loss is typically 5 dB maximum.

However, due to the power capability requirements of the combiners, a big problem is the size of the cavities, which are generally manually tuned to the desired frequency at installation. Any changes to the frequency will necessitate a visit to the BTS site in order to retune the cavity manually. Servo-tuned cavities are now being developed that allow remote frequency tuning if changes to the network configuration are required. This tuning is, however, not quick enough to support frequency hopping. The constraints for frequency hopping are dealt with in Section 7.7.2.5.

Hybrid Combiners

Hybrid combiners provide a passive combination of the transmitted signals. This kind of combiner is very lossy and provides very little selectivity in its output. The advantages of this type over the cavity are its size and wide bandwidth. It does not require tuning, so it can accommodate changes to network configurations and support conventional frequency hopping.

Problems arise because the combiner does not perform any filtering in the transmitter path. This places enormous constraints on the design of the filtering in order to achieve the required specification on intermodulation products. Combiner losses in the transmitter path can be as much as 7 dB when combining four transceivers in the BTS. Since the BTS power is specified at the input to the combiner, the use of this type of combiner will have an impact on the resulting network design.

7.7.2.5 Frequency Hopping

For efficient reuse of channels, frequency hopping is built into by the GSM system. The use of frequency hopping can also help to improve the performance of static or slow-moving subscribers due to cochannel interference caused during slow fades. The effective hopping rate is one hop per TDMA frame (i.e., 1/4.62 ms = 217 hops/s).

Note that the difference between the MS and the BTS in frequency hopping mode is that the MS uses only three out of the eight time slots available for receive, transmit, and monitor. The BTS, however, uses all eight time slots, since it must be capable of supporting eight MSs in one frame. The BTS must also be capable of both receiving and transmitting in all eight time slots

The guard space between time slots is only 30 μs. Tuning the synthesizer to different frequencies within 30 μs is technically very difficult, and a number of configurations for hopping at the BTS were proposed.

RF Hopping

This method of hopping requires agile transceivers, as in the mobile, except that two or three synthesizers are generally used. This allows one synthesizer to be tuned while the others are being used. The tuning time for each individual synthesizer is now a minimum of one time slot (about 0.5 ms).

The main disadvantage of this method is that a hybrid combiner must be used, since there needs to be non-frequency-selective signal combining. Note also that a continuous transmission of the BCCH channel is required. Therefore, a "fill-in" transmitter may be used in between the two hopping transmitters.

For BTS configurations with few (2 or 3) transceivers, RF hopping is a better solution than the baseband hopping alternative considered next.

Baseband Hopping

This is the alternative method of hopping when a larger number of transceivers are used in one BTS. Baseband hopping uses a fixed-frequency transceiver and multiplexes a number of baseband processing systems in order to use the appropriate transmitter for the defined hopping sequence. This avoids the need for wideband hybrid combiners because the resulting frequency of each transceiver is fixed, so selective cavity combiners can be used. By fixing the RF frequency, a continuous "fill-in" BCCH channel frequency can be provided. The disadvantage is that the configuration then requires one transceiver to be allocated for each frequency. It is therefore only cost-effective in large systems that already have a number of transceivers at the BTS.

Hybrid Hopping

Basically, this is a combination and compromise of the two methods just described. For the receive path, RF hopping is used because the need for wideband filters over the GSM frequency range does not present a problem for the selectivity of the BTS. For the transmit path, baseband hopping is used, so the output losses can therefore be minimized, and the intermodulation rejection requirements are met with the extra selectivity provided by the cavity combiners.

7.7.2.6 Receiver Performance

Another important feature of the BTS receiver is its ability to cope with space diversity in order to improve the uplink performance of the system (MS to BTS). In this situation, two antennas and two receive paths are used to process the direct and indirect signals. The resulting information after equalization is analyzed and compared, and the most accurate and highest quality signal is selected, or a combination of them both is constructed.

Due to the TDMA nature of GSM, a lower-cost alternative method is to use one equalizer to process both received input signals. However, subsequent delays may arise due to the amount of pre- and postanalysis required.

7.8 TIME SCALES AND TYPE APPROVAL

GSM was planned to take 4 years from conception to implementation. However, due to the complexity of the system and the fluidity of the recommendations, some issues took much longer than anticipated to resolve. Some of the technical requirements could not be confirmed until pilot systems were installed. Much investigation and testing were required, and it was not surprising that a number of implementation problems were identified. These had to be resolved to the satisfaction of all the parties and countries involved.

The GSM recommendations for Phase 1 GSM were finalized in January 1990, but issues are still arising as type approval is being performed. The Phase 1 recommendations have been updated and re-released. This process is likely to continue until full type approval is available.

Because of the fluidity of the recommendations, a major problem arose with the type approval of GSM terminals. This resulted in an interim type approval (ITA) being proposed so that the network operators could have approved mobiles for use on their networks. Once the ITA was agreed on, the GSM committee defined the major parameters that would need to be tested in order to ensure a satisfactory network operation until full type approval was available. From Recommendation 11.10, 160 ITA tests were identified.

The availability of full type approval testing was very much dependent on the provision of the system simulator and a full set of test scripts. Technical problems discovered during ITA led to new requirements being placed on the system simulator performance, but these issues have now been resolved. Full type approval is planned to be available by January 1994, when the license for the ITA expires. Phase 2 brings a new set of requirements for the system simulator, and these in turn will, no doubt, introduce a new set of problems.

7.9 CONCLUSIONS

A brief discussion of the aspects of the design features, constraints, and problems for GSM equipment designers has been presented, although this is, of course, by no means a complete list of the issues.

GSM is a complex system, and only now are its full implications being appreciated. The delay of product launches, type approval, and infrastructure rollout serve to give an indication of the technical challenges introduced by the specifications and the time scale demanded.

The challenge of GSM has certainly provided the opportunity to use large-scale integration and new design techniques. Overcoming the last few hurdles will lead the way to a European communication system and mass market, which should mitigate some of the development costs.

REFERENCES

Marley, N., *GSM & PCN Systems and Equipment,* JRC Conference, Harrogate, 1991.
Balston, D. M., *The Pan-European Cellular Technology,* IEE Conference Publication, 1988.
GSM/ETSI Recommendations, European Telecommunications Standards Institute, Sophia Antipolis, France. Specifically (version numbers are not given):
03.20 *Security Related Network Functions*
04.08 *Mobile Radio Interface—Layer 3 Specification*
05.05 *Radio Transmission and Reception*
05.08 *Radio Subsystem Link Control*
06.10 *GSM Full Rate Speech Transcoding*
11.10 *Mobile Radio Conformity Specification*
Mouly, M., and M.-B. Pautet, *The GSM System for Mobile Communications,* Palaiseau, France, 1992.

Chapter 8

The Birth of Personal Communications Networks

A. Hadden and P. Knight

Mercury Personal Communications[1]

The DCS 1800 version of GSM is being deployed in the United Kingdom and Germany in order to increase the competition between cellular operators. This chapter describes the background of DCS 1800, outlines the differences compared to GSM at 900 MHz, and discusses in particular the concept of the parallel network architecture, a scheme whereby two operators share a common infrastructure.

8.1 THE U.K. REGULATORY SCENE

8.1.1 Background

Two competing mobile communications networks began service in the United Kingdom in January 1985, operating according to the TACS specification (derived from the AMPS cellular standard) in the 900-MHz frequency band. Although the takeup of cellular telephones has been high (1,425,510 customers as of February 1, 1993 [1]), particularly towards the end of the last decade, the cost of service remains relatively expensive, which continues to limit its appeal. Furthermore, both operators faced considerable difficulties in matching their network capacities with the rapidly increasing demand, with a consequential decline in service quality. A solution was therefore sought that would satisfy demand for a high-quality, low-cost service, and yet be affordable to consumers in a mass market.

It was against this background that the U.K. Government's DTI decided to introduce more competition, not only to the cellular networks, but by using new radio-based techno-

---

[1]Now trading as Mercury One-2-One.

logies, it was envisaged that more effective competition could be created against the public fixed-link telecommunication services.

British Telecom (BT) continues to enjoy a dominant position in fixed-network services, and the licensing of a second operator in 1984—Mercury Communications, Ltd.—has resulted in BT losing only a small market share. The DTI initiative offered for the first time a prospect for a significant level of competition to BT, which radio delivery could be expected to achieve at lower cost than conventional wired systems.

In January 1989, the DTI published their consultative document *Phones on the Move* [2], which embodied a vision of a new kind of telephone service available through the use of a single personal handset connected to a PCN by radio. This would offer a national high-capacity two-way, fully mobile service competing with both existing cellular and public switched fixed-network services. In order to ensure that the proposed new networks would not suffer the capacity restrictions of the existing TACS networks, spectrum was reserved in the band 1,710 to 1,880 MHz.

The results of the DTI's international consultation process proved valuable in determining an appropriate vision of what PCN should be and how it should develop.

The DTI decided that, in view of its competition strategy, applications for licenses to operate the new PCN networks would not be considered from either the existing cellular network operators or BT, the dominant fixed-network operator. Instead, the DTI announced its intention to award a license to a consortium led by Mercury to allow them to compete more effectively across a wider range of services with BT, and simultaneously launched a competition for additional licenses. In December 1989, two further licenses for PCNs were awarded to Unitel and Microtel. Subsequently, in March 1992, the major shareholders of Mercury PCN and Unitel decided to merge their PCN interests into a single, new, jointly owned organization, known as Mercury Personal Communications.

8.1.2 Opportunities for Competition in the Local Loop

PCN systems in the United Kingdom will develop to provide nationwide coverage, but will initially offer ubiquitous coverage on a local or regional basis. In the medium term, the PCN service will evolve to address the local loop and thus provide a real alternative to conventional fixed-wire telecommunications services.

In economic terms, the copper local loop represents such a large investment historically that it is unrealistic to attempt competition requiring a similar scale of investment. Furthermore, the cost-per-subscriber connection through the use of radio is usually lower than in a wired system, especially when the costs such as laying cables and maintenance are taken into account. Thus, the use of radio in the local loop offers a viable alternative.

Opportunities in other countries can be exploited according to the various regulatory/ licensing strategies that can be contemplated, depending on national economic and political considerations. Areas not previously connected to the PSTN could benefit from PCN on a local, regional, or national basis, through individual, competing, or complementary networks, or be integrated into the existing PSTN.

8.1.3 Regulatory Initiatives

In recognition of the major costs of establishing the new PCN networks and in order to secure the economic conditions for mass market acceptance, the DTI proposed specific initiatives to lower the cost base.

- PCN operators are allowed from the outset to provide their own millimetric radio links between base station and sites and switching centers. The incumbent cellular operators were prevented from doing this until two years after the last of the PCN networks enter service. This restriction on the cellular operators has subsequently been refined and now prevents self-provision until April 1995.
- PCN operators are allowed to share infrastructure, which is especially useful for serving low traffic areas or rural areas where environmental and cost considerations are of major importance.

In order to ensure that PCN networks could support a mass market, the DTI reserved spectrum in the band 1,710 to 1,880 MHz (initially 2 × 25 MHz per PCN operator) for PCN services in the United Kingdom.

8.2 STANDARDS

The DTI recognized from the outset the importance of adopting internationally agreed-on standards for the ultimate success of PCN, including the conditions for roaming between PCN networks. It was stipulated that the specification must be based either on the emerging digital European cordless telephone (DECT) standard for cordless private branch exchange (PBX) or on the pan-European GSM digital cellular mobile system.

All PCN licensees stated their preference for developing a standard for PCN based on GSM, redefined for operation in the 1,710- to 1,880-MHz band, and incorporating initial PCN requirements. This view was consistent with one of the key recommendations made by the ETSI Strategic Review Committee on Mobile Communications in Recommendation 8—Digital Cellular System at 1,800 MHz (DCS 1800).

> GSM must be asked to elaborate an enhancement to the GSM standard for using frequency bands compatible with the plans of CEPT and located in the vicinity of 1.8 GHz. This new version of the standard should be aimed primarily at providing a service for handheld or pocket terminals in densely populated zones and must be suitable for pan-European implementation.

GSM was chosen as the basis for PCN because it provides the best match for PCN requirements, defining a complete system incorporating:

- Radio, transmission, and switching networks;
- Network, radio link management, and mobility management;
- Feature-rich set of services;
- Potential for economic wide-area coverage;

- Open interfaces;
- Potential for use in the local loop.

This recommendation was accepted by the ETSI Technical Assembly in March 1990, and thus DCS 1800 was developed as the European standard to meet the requirements for PCN.

8.3 REQUIREMENTS AND MARKETING OPPORTUNITIES

The term *personal communications* has been adopted by several operators seeking to align their respective service offering with a "personal" concept. Personal communications is defined as a service that fully meets individual needs, combining the attributes of available technologies—cellular, cordless, paging, and so on to deliver a "personal" service to customers using a single handset terminal device anywhere within the coverage area.

This is the PCN concept in the United Kingdom, which focuses on the provision of high-quality two-way communications services—speech and data—to business users and consumers on the move outdoors and indoors. The goal in addressing the mass market is such that, in time, PCN will be used in the home as a competitive alternative to the conventional public switched network service. PCN must therefore find solutions to the problems of delivering the required range of speech, data, and supplementary services at a high quality level from a competitive cost base to realize a high-capacity service for portable, in-car, and in-building usage.

A key requirement of PCN is its ability to support pocketable, lightweight, easy-to-use, low-cost handsets that will be carried anywhere. The choice of power class and radio specification parameters must therefore allow close-proximity working between handsets and serving/nonserving base stations, and support attractive standby and talk time performance in order to be consistent with supporting a mass population of PCN subscribers. Furthermore, the technical standards for PCN need to support a multioperator environment, since PCN services will be implemented in a highly competitive market in many countries, commencing in the United Kingdom and followed by Germany.

Customers need to be able to make calls to and receive calls from anywhere in the world, and service packages will need to align with the telecommunications needs of specific user groups. These features will give customers full control over their calls; for example, customers will be allowed to divert calls to a voice mail service and access them at a more convenient time.

Studies showed a broad recognition of the benefits of personal communications in both business and residential sectors, and a sense of inevitability that personal communications will become part of everyday life. They also indicate that significant changes in work patterns, attitudes, and lifestyles will take place in the 1990s, which will result in increased demand for personal communications services. These factors, when coupled with quantitative demand studies, indicate a total mobile services market size in the region of 7 to 10 million mobile services subscribers in the United Kingdom by the year 2000 (see Figure 8.1). For Western Europe, the combined PCN/Cellular subscriber base has been estimated to be 17.88 million in 2000, rising to over 30 million by 2010, as shown in Figure 8.2.

These trends towards mobility demands and usage can be expected to apply similarly to other well developed economies. PCN thus represents what is probably the most significant opportunity in mobile communications development over the next 10 to 15 years.

8.3.1 Comparison of Cellular and PCN

The technology on which both GSM and PCN networks will be based is cellular radio. There is, however, a clear differential in how GSM and PCN operators will exploit the market opportunities. Whereas the "classic" cellular (AMPS, TACS, NMT, etc.) service is optimized primarily for high-power mobile (vehicle-installed) terminals, the emphasis in PCN is on access to the network for small, lightweight "go-anywhere" handsets. Furthermore, since PCN must be relied on to offer service throughout the coverage area, attention is focused on finding solutions for consistent in-building service. Finally, with its emphasis on the mass market, PCN needs to be affordable by ordinary people, who can be expected to weigh the value for money of the mobility that PCN provides against the services provided by the fixed PSTN.

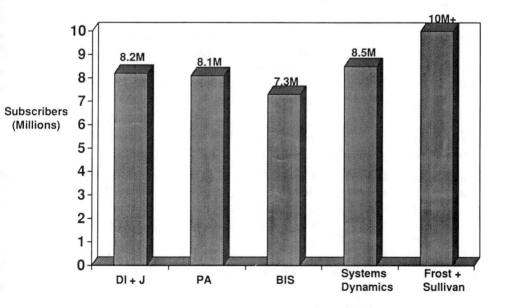

Figure 8.1 Forecasted demand for mobile voice services in the U.K. in the year 2000.

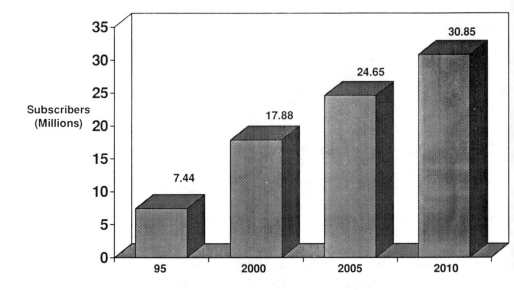

Figure 8.2 Growth in demand for mobile services in Western Europe: PCN and cellular subscriber base.

8.4 DCS 1800—INITIAL SPECIFICATION

In view of the early launch date, changes to the GSM standard for DCS 1800 were deliberately restricted to necessary modifications in the RF area (requiring adaptation from 900 MHz for operation in the 1,710- to 1,880-MHz bands), the associated signaling modifications, and to providing the technical capabilities for infrastructure sharing.

The frequency of operation of DCS 1800 is twice that of GSM, the exact bands being:

- Mobile transmit, 1,710 to 1,785 MHz (GSM 890 to 915 MHz)
- Mobile receive, 1,805 to 1,880 MHz (GSM 935 to 960 MHz)

The carrier spacing of DCS 1800 is 200 kHz, as it is for GSM 900. The absolute frequency of any given channel is given for DCS 1800: $1710.2 + (0.2) \times (n - 512)$ MHz, where $512 \leq n \leq 885$ and n is the absolute radio frequency channel number (ARFCN).

Using the above notation allows the carriers defined for DCS 1800 and for GSM 900 to be uniquely referenced.

The GSM 900 system supports 124 carriers, which are encoded to provide the mobile with information about which frequencies are used when frequency hopping, the list of BCCH carriers to be used in idle mode for cell selection, and for measurement reporting activities. Changes were needed for DCS 1800, which is required to support a total of 374 channels.

Recognizing as a common goal that Phase 2 developments for both DCS 1800 and GSM should become aligned as far as possible to encourage efficiencies through consistent designs, and also bearing in mind that GSM itself could be expected to expand into extension bands, it was decided that a new encoding scheme should be developed to support 1,024 channels.

Changes were needed too on the SIM card, where the current standard stores BCCH carrier information as a 124-bit bit map. To ensure compatibility between GSM and DCS 1800, a new DCS 1800 application and directory was defined, as shown in Figure 8.3. This would allow a mobile to be able to recognize the DCS 1800- or GSM-related information as appropriate. The SIM data field structures and application protocols used for DCS 1800 are identical to those existing for GSM.

It is possible to envisage that SIM cards could be produced containing both GSM and DCS 1800 applications, since the additional memory overhead is small, being of the order of 165 bytes only. The use of such a dual-application SIM card can be foreseen for a roaming service between cooperating GSM and DCS 1800 networks, where, for example, the DCS 1800 customers insert their SIM cards into GSM 900 mobile equipment, and vice versa.

A key requirement of PCN is its ability to support small, low-cost handsets. Target operation range scenarios were chosen at 8 km (rural) and 1 km (urban). A 1W handset was specified, together with an alternative of 250 mW, *both significantly* below the power classes that will typically be used in GSM 900 MHz systems. Considerable attention has been paid to the choice of radio parameters to allow close-proximity working between handsets and base stations, since widespread adoption of PCN services is expected. Fundamentally, GSM was designed with 20W or 8W car mobiles in mind, and the presence

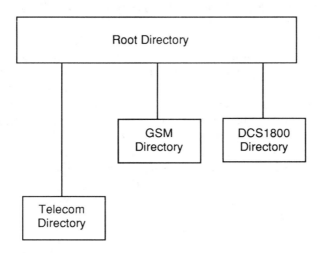

Figure 8.3 SIM card directory structure.

of such high-power transmitters in GSM networks will inevitably limit capacity due to sharp falloff in service quality in their vicinity. The four lowest base station power classes from GSM 900 were retained for DCS 1800.

The methodology used to develop the DCS 1800 standard was to identify a number of scenarios considered relevant to actual deployments, and after agreement on basic working assumptions (e.g., antenna heights, path loss (free-space and in-building)), to decide on radio specification parameters that would support PCN service and implementation requirements. The following scenarios were evaluated and treated in this way.

- Single BTS and mobile;
- Multiple mobiles and BTSs, assuming they belong to the same network (i.e., coordinated);
- Multiple mobiles and BTSs, belonging to different networks;
- Collocated mobiles served by different networks;
- Two or more collocated BTSs from different networks;
- Collocation of DCS 1800 mobiles in close proximity with other systems, such as GSM, DECT, CT2, and analog cellular systems (including TACS, NMT 450/900, AMPS, C-450, R2000).

Working in this way, it was possible to set the detailed RF parameters, such as transmitter characteristics (for the handsets and base stations), and receiver characteristics (blocking, intermodulation, and spurious emissions). The composition of the ETSI technical committees, which includes manufacturers and operators, ensured that the final specification represents a fair balance between requirements and what can be produced consistent with cost objectives. This methodology of considering operational scenarios proved to be a very beneficial work approach, and will probably be employed more extensively in other telecommunications standards-setting activities.

A key enhancement of the specification for DCS 1800 has been the incorporation of national roaming, which allows roaming between overlapping networks within a country, whereas GSM is designed for international roaming between nonoverlapping areas of coverage. The technical capability is provided for subscribers to automatically obtain service from other PCN networks if coverage is not available from their own network and to return automatically to their own network when coverage is available. This capability will be particularly useful for coverage in rural areas where there are environmental sensitivities, and will also effectively accelerate service rollout in the early years. DCS 1800 additionally supports international roaming. The possibility of sharing physical equipment as an enhanced form of national roaming is currently being studied.

The main requirements for national roaming sought by the PCN operators are summarized as follows, assuming commercial agreements are in place between operators wishing to offer a national roaming service:

- The roaming service would be available on a location area basis rather than on a whole PLMN basis, which GSM currently specifies.

- There would be automatic selection of another PLMN when the mobile is outside home PLMN (HPLMN) coverage.
- There would be automatic return to the HPLMN when within coverage and after rejection of a location area update request.
- Any network offering national roaming must additionally support the (GSM standard) international roaming service to customers of other networks.

The requirements represented a considerable amount of technical work, and changes were required to six specifications, including the type approval test specification 11.10. The automatic selection of the PLMN feature to meet the above requirements was foreseen through preprogramming. It was decided to include a new reject cause, "national roaming not allowed in this Location Area," following a location area update request from the mobile. In order to avoid possible service degradation, selection of the HPLMN and invocation of the cell selection procedure were chosen for selecting a new cell.

The main features of national roaming are presented in Figure 8.4.

A major design concern was to avoid a possible ping-pong effect that could occur from rejecting a mobile trying to access two forbidden location areas. A solution to this was found by including in the mobile a list of forbidden location areas—which is updated if the new reject cause described earlier is received. This list is erased at power-off, on removal of the SIM card, and periodically during the idle mode.

National roaming was not specified for Phase 1 of the GSM 900-MHz standard, being developed initially for DCS 1800 applications. However, this capability was included for GSM in its Phase 2 specification development. This raised the question of upward compatibility, since a GSM network implemented in the Phase 2 specifications including the national roaming capability must ensure that Phase 1 mobiles can be handled correctly. Similarly, Phase 2 mobiles must be capable of obtaining Phase 1 services when operated within a Phase 1 network. These objectives are met. A further problem was that which could occur at country borders, and the need to ensure that mobiles do not select a foreign

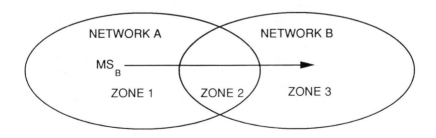

NATIONAL ROAMING

Figure 8.4 The main features of national roaming: Zone 1, MS_B served by visited network A; Zone 2, MS_B automatically returns home to network B; Zone 3, MS_B served by home network B.

network while within coverage of a national network offering national roaming. It was decided that networks offering national roaming should be given a higher priority in the list of preferred PLMNs contained in the SIM.

The DCS 1800 standard was released by ETSI as a series of delta specifications that detail the deviations and enhancements of the core GSM specifications. See Table 8.1.

Since DCS 1800 is based on GSM, PCN operators will be able to exploit a powerful signaling system between the handset and the network, which has similarities to the ISDN. However, for DCS 1800, the frequency encoding used to broadcast lists of RF channels for network access required adaptation, due to the increased number of channels.

PCN operators will also exploit the SIM already specified for GSM. The SIM—which for handsets may be full-size (ISO) or plug-in, depending on preferred product implementations—will contain information about the customers and services to which they have subscribed, billing details, and mechanisms relating to the caller authentication process. The SIM specifications are amended to permit a DCS 1800 directory and other data to be stored on the SIM instead of or in addition to GSM fields. The technical capability exists for DCS 1800 customers to benefit from a roaming service to a GSM 900 network by means of the SIM, and vice versa. These developments are the forerunners of multiapplication Smart Cards currently being studied in numerous standardization bodies in Europe and elsewhere.

Table 8.1
Specification Changes Compared With GSM

Specification	Details of Change
02.06-DCS	Definition of the two DCS 1800 handset power classes
02.11-DCS	Definition of national roaming
03.12-DCS	Modification of location registration procedure for national roaming
04.08-DCS	New procedures in the handset for national roaming; changes to the air interface messages to accommodate the large increase of channels for DCS 1800
05.01-DCS	Definition of DCS 1800 frequency band and performance
05.05-DCS	Substantial changes to the core GSM specification to define the changes in handset and base station radio performance and especially to allow for a high population density of handsets
05.08-DCS	Modifications of the link access between handset and base station to allow for the substantial increase in channels for DCS 1800 and to provide the national roaming facility
08.58-DCS	New messages on the BSC-BTS link arising from the increase in number of channels
09.02-DCS	Changes to the mobile application point (MAP) to provide national roaming
09.10-DCS	Definition of conversion of messages between the MAP and air interface
11.10-DCS	DCS 1800 Handset Conformity Specification
11.11-DCS	Specification of DCS 1800 directory on the SIM; changes to the SIM to allow for the increase in number of channels
11.20-DCS	DCS 1800 Base Station Specification
11.40-DCS	DCS 1800 System Simulator Conformity Specification

First implementations of PCNs [3] will use the DCS 1800 standard, with the world's first commercial PCN service expected to be offered by Mercury Personal Communications during the summer of 1993.

8.4.1 Phase 2 Standards

A common evolution of the DCS 1800 and GSM 900 specifications was agreed on as desirable. The action plan for Phase 2 includes a task to combine the core recommendations for PCN and GSM into a single set of documents for administrative benefit. The principal activities within the Phase 2 work program for release during 1993/4 are:

- Additional services/features to ensure competitiveness with present day analog systems;
- Optimization in the light of operational experience;
- Adaptation to allow export to non-European countries;
- Microcell enabling techniques;
- Enhancements to national roaming;
- Enhancements to short-message service.

See Table 8.2.

Table 8.2
GSM 900/DCS 1800 Standards—Minimum Technical Capabilities

Phase 1	*Phase 2*
Voice (full rate 13 Kbps)	Signaling/multiplexing for half-rate channels*
Data async/synchronous services up to 9.6 Kbps PAD access up to 9.6 Kbps async Async data packet up to 9.6 Kbps	Half-rate data channels
Short-message services	Enhancements to SMS, including memory free, delivery confirmation
Supplementary services, including call barring and call forwarding	Additional supplementary services, including call waiting/call hold, multiparty, line identification, and closed user group
Group 3 facsimile (transparent and nontransparent services)	
National roaming (DCS 1800 only)	Enhancements to operational characteristics of national roaming
	Microcell enabling techniques
	Operator-determined call barring
International Roaming	

* The half-rate codec specification is phase-independent.

8.4.2 Microcell Techniques

Microcells will allow future capacity enhancements in PCN networks without requiring substantial modification to the existing network. Microcells are defined as small cells whose base station transmitter is placed below rooftop height so that RF propagation is confined to a small local area. This allows the reuse of RF carriers at smaller distances, giving increased network capacity. The Phase 2 standard will eliminate the problem of high-speed mobiles dropping calls when they leave the coverage area of the microcell due to their occupying the microcell for less time than is required to initiate a handover. The solution will encourage slow-moving mobiles to access the microcell, while ensuring that fast-moving mobiles remain served by macrocells.

The technique of encouraging mobiles to select certain cells in preference to others is known as *cell prioritization.* An operator may seek to do this for several reasons. For example, consider a scenario whereby small towns along a main road are each served by small cells, with a larger "umbrella" cell deployed to provide coverage between towns. Since the umbrella cell is covering a larger area, higher transmitter power is used, which may cause a mobile to select that cell, even if it is within coverage of a smaller cell. Once a call is made, the network may initiate cell handovers to other small cells, which again could be within the umbrella cell. These handovers lead to a network signaling load, which could be avoided by ensuring that the mobile always selects the small cell when available and does not access the umbrella cell.

In other situations, the operator will be concerned with achieving high-capacity performance through the use of layered cells; a pattern of normal cells are deployed to provide contiguous coverage, while microcells provide additional capacity in targeted areas. Demands on the network are also affected by factors such as time of day (e.g., commuting patterns). Although the microcell provides capacity where it is needed, there are limitations in handover performance due to its small area; the speed at which a mobile could cross the microcell may be faster than the handover algorithm can support. The solution adopted is to ensure that a fast-moving mobile does not select a microcell when the larger cell is available. This is accomplished by the handset delaying its selection of a microcell for a period determined by the network operator. If the microcell is still present after this period, then it is selected. Although this does not totally prevent fast-moving handsets from selecting a microcell, it does mean that they are unlikely to select it until they are well within coverage of the microcell, which will give the handover algorithm sufficient time to take the appropriate action following any subsequent call setup. The technique allows sufficient differentiation between slow-moving users and, for example, vehicular-based users, the latter being "steered" towards the larger cells.

It is also possible for an operator to restrict mobiles from accessing certain specific cells, which may be necessary for some operational requirements, perhaps when reconfiguring cells or when adding new cells to the network. Information about which cells are barred is detected by the mobile from regularly updated transmissions on the BCCH channel.

8.5 THE PARALLEL NETWORK ARCHITECTURE CONCEPT

The construction of a new mobile telecommunications network using DCS 1800 technology represents a major capital investment project. Utilization and sharing of existing or emerging telecommunications infrastructures represents a potential opportunity to reduce investment and risk, and in the United Kingdom this was explored through the use of the parallel network architecture (PNA) concept, between Unitel and Mercury PCN, prior to the merger that created Mercury Personal Communications. The principal benefit of adopting PNA is that it reduces costs that, in the United Kingdom, were expected to realize savings of up to 40% for each operator. The PNA operation was established as a company jointly owned by each operator.

PNA exploits the intelligence implicit in a mobile network to segment a common physical network into multiple logical networks. In essence, PNA allows two operators to share physical transit infrastructure elements, with each operator managing its own customers and services. Customers will "see" only their operator, and can therefore choose between the services and features offered, together with different customer care, billing, and tariffing arrangements. Therefore, adopting PNA does not diminish competition, since each operator is free to offer its own services. Alternatively, both operators may choose to offer similar services, but with differing marketing/tariffing propositions.

8.5.1 Maintaining Competition

Before the merger, a great deal of work was done towards the realization of PNA, which can offer regulators an attractive solution to lowering the cost of entry for new operators in a competitive environment. The need to allow two or more companies implementing PNA to compete effectively is an important factor, and some explanation is required about how this can be achieved. The key aspects of competition are:

- Radio coverage and quality;
- Interconnect dimensions and specifications;
- Core network services;
- Value-added services;
- Administration system functions;
- Marketing strategies and branding;
- Distribution and sales channels.

8.5.2 Coverage and Quality

Coverage and service quality are the principal factors differentiating between network operators as perceived by the customers. The sharing of base station sites and equipment implies identical measures for cooperating operators. However, differentiation can be achieved where each operator has the opportunity to implement unique cells to optimize individual network operator requirements either inside or outside the shared area. Such additional cells would of course be 100% funded by that operator.

8.5.3 Interconnect

The specification of interconnect requirements with other networks remains the responsibility of individual operators, thus allowing agreements to be reached that can be tailored to support specific business opportunities and requirements.

8.5.4 Core Network Services

In a traditional cellular network, the majority of the services are resident in MSCs, or their equivalent, and are built into the cellular switch application code. This means that each customer is offered a subset of the total service set according to the subscription options that have been chosen.

In a PNA environment, this could restrict competition unless other techniques are found to offer differentiated services. Two techniques were explored:

- The exclusion of the switching centers from the PNA agreement;
- The use of additional intelligent network nodes to provide additional services.

Exclusion of the MSCs from the PNA agreement would have allowed the two PNA operators to use a common set of base sites and individual switches. However, the range of services available were generally defined by GSM in great detail anyway, and therefore on its own, the simple exclusion of the MSCs from PNA was not seen as an effective means of allowing competition.

The use of IN techniques, however, allows each operator to define its own services in addition to the basic set of services already available in the GSM standards. The flexibility of services that can be offered through this technique is then only limited by the sophistication of the IN platform and the investment that each of the operators is prepared to make. The PNA operators therefore concluded that the use of IN techniques was the most effective means of allowing them to individually improve upon the GSM basic set of services and to thus effectively compete with each other and the other cellular operators in the United Kingdom.

8.5.4.1 Value-Added Services

Value-added services (e.g., voice messaging, SMS) are an important area where traditionally there has been a considerable amount of difference in the various options marketed by different cellular operators. It was clear that the platforms used to support such services should be excluded from the PNA agreement and that each operator should provide its own capabilities.

8.5.4.2 *Administration System Functions*

The administration functions of a network are often overlooked in the analysis of the overall network. However, in practice they are a source of major differentiation between the operators. Administration functions cover the following key areas:

- Billing of customers;
- Handling general customer inquiries;
- Allowing customers to register for service with the network;
- Handling any stocks that the operator may hold (e.g., handsets);
- Operating core network and value-added systems.

A major source of differentiation is the tariffing structure. This can include a level of handset subsidy and tradeoffs between monthly service and call charges.

Ultimately, the billing computer is the device that affords operators flexibility in the tariff regime they wish to implement. For this reason, it was decided that the billing systems should not be included in the PNA core system and that each operator should implement its own system.

The "customer care" relationship between the operator and the customer is also viewed as an area of differentiation between operators. There is an enormous range of operator-controlled services that can be implemented and that are very visible to the end user. In addition, the inquiry bureau of an operator will often be the point of contact for customers that are experiencing problems with their service, and they will judge the operator by the attitude with which they are treated and the efficiency with which their complaint is handled. For these reasons, the inquiry desk functions were excluded from the PNA core.

8.6 IMPLEMENTATION ISSUES ASSOCIATED WITH PNA

The resultant architecture for the U.K. PNA scheme is shown in Figure 8.5. Having set this architecture, there are then several other issues to be addressed concerning the operation of the network.

8.6.1 Allocation of Transceivers

Transceivers at the base sites of a PNA system provide up to eight full-rate channels each. The amount of traffic that can be carried by a pool of channels is measured in erlangs as described in Chapter 1 and is a nonlinear function such that the efficiencies of any given pool of traffic channels are increased as the size of the pool increases for a given quality of service.

In the derivation of the PNA architecture, an issue that caused much debate concerned how transceivers at any given base site should be allowed to carry traffic for each of the two competing operators. There were two broad options.

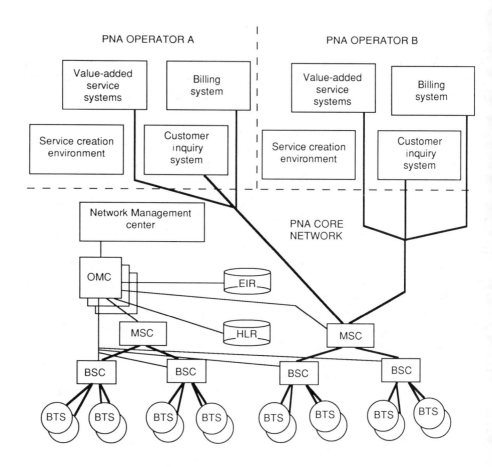

Figure 8.5 Parallel network architecture.

1. The traffic channels are pooled and offered on a first-come first-served basis to either operator's customers.
2. Each transceiver is assigned either to operator A or operator B and is used exclusively for that operator's customers. This means that at each PNA base site there are two pools of transceivers, each of which operates at a lower level of efficiency. The total traffic carried is therefore less than that in option 1 for the same number of transceivers at the base site.

The advantage of option 1 is that the transceivers are used efficiently; the disadvantage is that the amount of competition between the two PNA operators is limited.

In the United Kingdom, a compromise solution was adopted such that transceivers were allocated to individual PNA operators, but in order to achieve the additional efficiencies of a pooled set of transceivers, an overflow of traffic was allowed between the two pools of transceivers, and this is accounted for by a commercial agreement between the two PNA operators.

This is an area where there are considerable savings in infrastructure costs to be made, but only at the expense of competition between the PNA operators. How this balance is achieved becomes an issue for the regulatory regime of the country allowing the employment of PNA techniques.

8.6.2 Billing

Traditionally, the billing system takes input from the switching system in the form of call data records. These are files of information that contain details of each billable call; for instance, who made the call, where the call was originated, where and how it was terminated, and the duration and absolute time of the call.

Each record can be quite large, on the order of 100 to 200 bytes of information, and for a large network, this amounts to a considerable amount of data to be passed around the network.

The use of a PNA system causes an additional complication in that the call data records dealing with each of the operator's customers must be streamed to the correct billing system. Two approaches can be adopted to achieve this: either streaming directly on the switches and then generating two billing streams, one to each billing system; or employing an adjunct processor, which takes a single stream from the switch and splits this stream into two by analyzing the identity of each mobile customer. Both approaches were found to be equally acceptable, and the former option was preferred in the United Kingdom because it resulted in lower overall project risk.

8.6.3 Roaming

Roaming is the ability of a mobile from one network to visit another network and obtain service from that network by updating back to the parent HLR. This clearly requires an agreement between the home and visited network.

The use of a PNA technique allowing two operators to share a core network raises the issue of how to deal with mobiles that roam onto that network. The service given to these mobiles has to be billed to the home network via one of the two PNA operators. This in turn complicates the call data records' streaming algorithms within the network, which have to stream to the appropriate PNA operator for mobiles that may come from many different other operators.

8.6.4 Customer Data Administration

The customer administration data are held in a database referred to as an HLR, as explained in Chapter 1. In a non-PNA network, these data are administered by the operator's customer administration system. In a PNA network, each operator has its own administration system that gains access to data in the HLR and modifies it. Precautions have to be taken to ensure that each operator's administration systems only have access to areas of the HLR(s) that contain information for that operator's customers. Clearly, it would be unacceptable to allow competing operators to either access or change the service entitlements of their competitor's customers.

8.6.5 Distribution Marketing and Sales

The sales channels are not included in the PNA relationship. This allows the two operators to sell and market their services in whatever way they feel appropriate, thus maintaining competition.

8.6.6 Planning and Design of the Network

The planning and design of the elements of the network included in the PNA agreement are undertaken by staff involved in the joint venture company. Any elements remaining in the parent companies (for instance, the billing systems and value-added service systems) are planned and designed by the parent company involved.

8.7 CONCLUSIONS

The PNA approach inevitably complicates network operation because activities are split between three operational groups (i.e., the PNA operator and the two parent companies who own the customers). However, with a well-defined functional division between the three groups, the majority of the operational effort would remain in the PNA organization.

The merger between Mercury Personal Communications Networks and Unitel rendered unnecessary the implementation of the PNA concept. Sufficient analysis and planning was undertaken, however, to suggest that the model could prove a viable mechanism for establishing a degree of competition without the need for investment in two or more complete networks. This approach might prove particularly effective in the emerging countries of Eastern Europe.

REFERENCES

[1] Mobile Communications Financial Times, Business Enterprises Ltd., London, Issue 20, 25 February 1993.
[2] *Phones on the Move,* U.K. DTI Consultative Document, January 1989.
[3] Potter, A. R., "Implementation of PCN's Using DCS1800," *IEEE Communications Magazine,* December 1992.
[4] Gaskell, P. S., "Developing Technologies for Personal Communications Networks," *IEEE Electronics and Communications Engineering Journal,* Volume 4, April 1992.

Chapter 9

Introduction of Digital Cellular Systems in North America

F. Lindell and K. Raith

Ericsson Radio Systems AB

The regulatory and frequency management situation in the United States has put constraints on the move to digital cellular systems and prevented the "clean sheet" approach that was the hallmark of GSM in Europe. This chapter describes the development of the digital standard in the United States, outlines the key properties of the radio part, and looks ahead to future developments.

9.1 INTRODUCTION

The rapid growth of cellular systems around the world is expected to continue during the 1990s and to benefit from the introduction of digital radio technology. The equipment will then become cheaper and more compact. In Europe, with its multitude of analog cellular standards (see, for example, Chapters 3 and 4), the GSM standard has been chosen as the unified pan-European basis for the digital cellular systems of the 1990s. GSM will make roaming possible throughout Europe, and manufacturing volumes will be large. The GSM standard is not compatible with existing analog standards.

The situation is different in North America. One single analog standard has been accepted and roaming is made possible throughout North and Latin America, Oceania, and Asia, where several countries have accepted the AMPS standard (see Chapter 2). The size of the market, the economies of scale, and the stiff competition have resulted in AMPS becoming the most widely used cellular standard in the world. However, no new frequency band was available for a new, more spectrally efficient digital system, and those frequencies that were in use were allocated between operators in a piecemeal fashion.

This lack of contiguity in the frequency use was an important constraint on the type of solution that could be adopted in North America.

9.1.1 Driving Forces Towards a Digital Cellular System

The spectacular growth of the number of subscribers of cellular systems must be accommodated through a continual increase of system capacity, usually by reducing cell areas and introducing additional base stations. In most large cities, however, it has become increasingly difficult and costly to obtain the necessary permits to erect base stations and antennas. Network operators therefore wanted a solution that made it possible to increase system capacity significantly without requiring more base stations.

The FCC is an organization appointed by the U.S. Government to regulate the radio and telecommunications industries. In the cellular radio field, the FCC allocates frequency bands, assigns frequencies to operators, and sets the permitted levels of radiation into other frequency bands. Thus, the operators can use virtually any radio technology as long as they stay within the limits set for radiated power. The transition to digital cellular radio does not require any action by the FCC.

The TIA was formed in 1988 as the result of a merger of the U.S. Telephone Suppliers Association and the information technology group of the EIA. The TIA and EIA collaborate closely and both produce standards. For EIA and TIA standards to become U.S. standards, they must be approved by the American National Standards Institute (ANSI). Although these standards are not formally binding, they are in reality closely adhered to within the industry.

The CTIA represents the cellular operators in North America. CTIA formulates requirements for cellular standards, which are then produced by the TIA. The CTIA, as well as individual operators, attends TIA meetings and provides the TIA with information required for standardization.

It was obvious that the solution was to introduce digital radio technology. In March 1988, the TIA set up a subcommittee, TR-45.3, to produce a digital cellular standard. Six months later the Cellular Telecommunications Industry Association (CTIA) presented a requirements specification. The two organizations agreed on a time schedule, with standards being issued in steps. The first step would ensure a large increase in capacity through the introduction of digital voice channels. Future steps would be devoted to additional features and further increases in capacity.

The TIA network architecture model is shown in Figure 9.1. Subcommittee TR-45.3 specified only the Um interface. Most of the other interfaces are specified by TR-45.2.

- Mobile station—Contains the interface equipment needed to terminate the radio channel at the user, functions for speech communication, and interface for connecting data terminals;
- Base station—The radio equipment at a site, serving one or more cells;

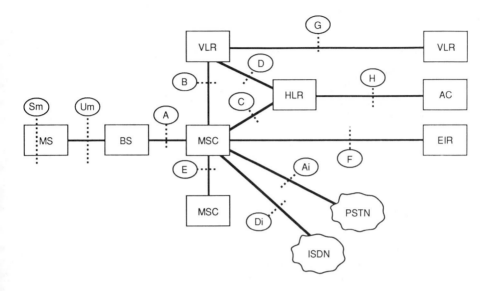

Figure 9.1 The architecture of the TIA network.

- Mobile services switching center—The interface for the user traffic between the mobile network and other public switched networks or other MSCs in the same or other mobile telephone networks;
- Home location register—The register of subscribers, specifying required services;
- Visited location register—A register separate from HLR, used by an MSC to obtain the information needed; for example, to handle a call to or from a roaming user who is temporarily within the coverage area of the MSC;
- Equipment identity register—Records the identity of mobile equipment;
- Authentication center—A unit that checks whether a caller is authorized to use the requested service.

One of the prerequisites of the standardization was dual-mode telephones (mobile stations); that is, they should be capable of operating on both analog and digital voice channels. This would allow network operators to introduce digital radio channels in city centers and other areas where capacity limits have been reached. Mobile stations would automatically switch over to analog channels if no digital channels were available. The subscribers would experience the same level of coverage as with the present analog systems.

The TIA requested proposals for digital systems from the member companies. When the proposals were presented, it became apparent that the main differences between them was the access method. Should FDMA or TDMA be used? Discussions centered on this question for several months.

9.2 THE CHOICE OF TDMA

TIA decisions are normally based on consensus. Different technical solutions are discussed until the members can agree on one. However, in this case neither the advocates of FDMA nor those of TDMA would give in. The matter was therefore settled by ballot. The result was that TDMA was chosen (by a large majority). The main arguments in favor of the TDMA method were that it permits:

- Easy transition to the new system;
- A handover procedure assisted by the mobile station; and
- Flexible user data rate.

9.2.1 Easy Transition to TDMA

In view of the large investment in the analog infrastructure, it was important that transition to the digital system be as simple as possible. With TDMA, it is possible to replace a 30-kHz analog channel with a digital channel having the same bandwidth. The digital channel can transmit simultaneous calls, and the frequency plan for the analog system can be retained. The combiner filters, in which the signals from the power amplifiers in the base stations are combined and then fed to the antenna, can also be retained. This would not be possible if the 30-kHz analog channel were to be replaced by three 10-kHz FDMA channels. FDMA would necessitate a reassignment of frequencies and filters in the system or the introduction of complex power amplifiers for both base stations and mobile stations.

9.2.2 Mobile-Assisted Handover

The anticipated rapid growth in subscriber numbers in conjunction with smaller cell sizes makes it increasingly important to be able to locate mobile stations faster and more accurately than in present systems. TDMA makes it relatively easy for a mobile unit to measure the signal strength on channels from neighboring base stations and report them to its current base station. The fixed part of the network—base stations and mobile-services switching centers—evaluates these measurements and indicates the base station to which the mobile unit will be handed over when it is about to leave a cell (or when, for any other reason, it would gain in radio link quality by a handover). The number of handovers increases when the traffic per cell increases, and the cell size is reduced. If the analog system method were used—where neighboring base stations measure the signal transmitted from a mobile unit—the signaling load on the links between the base stations and the MSCs would be very large, and this would also require very high data processing capacity in the MSCs. A decentralized location procedure, where each mobile unit is a measurement point, will thus reduce the load on the network.

9.2.3 Flexible User Data Rate

Future increases in network capacity require that the standard permits exploitation of developments in the field of speech coding, developments that continually reduce the bit rate required by the codec in order to maintain a given speech quality. If the bit rate from the speech codec is reduced by half, the capacity increases approximately to the same degree. With FDMA, the bandwidth of the radio channel must be changed in step with the bandwidth required by the user. This requires very narrow receive filters (5 kHz) and a stringent receiver specification if it is to be possible to introduce half-rate speech channels in an FDMA system. If different bit rates have to be used (e.g., different data rates and speech), the mobile units must be equipped with a switchable receiver. This is not a realistic solution for a small, low-cost, handheld mobile.

With TDMA, different users may use different data rates; they are simply given the time required (i.e., half the number of time slots for half-rate channels). This does not affect the radio part in the transceiver and, hence, does not increase the complexity of the telephones. The upper limit of the data rate offered by the system—if its complexity is not to increase drastically—is determined by the nominal channel bandwidth. In the North American system, the difference between TDMA and FDMA is at least a factor of three in favor of TDMA.

9.2.4 An Experimental System

Many people considered that a change from the FDMA method used in present-day analog systems to the TDMA method would constitute natural progress. There was concern, however, as to whether the time was right or whether TDMA needed more time to mature. An experimental system was built by Ericsson to demonstrate the feasibility of TDMA. A TDMA transceiver was connected to an operating base station belonging to the Los Angeles Cellular Telephone Company. Mobile stations, TDMA and analog reference sets, were installed in a vehicle, which toured urban and semirural areas (Figure 9.2). Most observers taking part in the demonstration judged the speech quality of the digital solution to be as good as, or better than, that of the analog method.

9.3 OTHER SPECIFICATION ISSUES

9.3.1 Speech Codec

One of the most important factors in determining the capacity of a cellular system is the bit rate of the speech codec. With a low bit rate, the amount of spectrum-time consumed for one connection is low, permitting more simultaneous connections within the system bandwidth.

Figure 9.2 Ericsson built an experimental system in order to demonstrate the advantages of the TDMA method. A TDMA transceiver was connected to a base station already in operation, belonging to the Los Angeles Cellular Telephone Company. TDMA units and analog units were installed in a van that toured urban and less densely populated areas.

The method used to choose the algorithm for the speech coder was to test hardware from nine suppliers. The test included sensitivity-to-bit errors. The candidates could allocate a part of the bit stream to error correction. The total bit rate should not exceed 13 Kbps.

The candidates were tested under various conditions, and the results were evaluated through a subjective listening test in which 100 persons participated. The winning speech and channel codec algorithm used 8 Kbps for speech coding and 5 Kbps for error detection and correction.

This was the only case where hardware testing was used. All other decisions were made by the committee. The speech codec chosen uses a variant of the code excited linear predictive (CELP) algorithm, and protection against bit errors is provided by a cyclic redundancy check (CRC) error detection code followed by a convolutional code with a constraint length equal to 6. The total complexity for a duplex implementation is about 15 to 20 MIPS. The subjective quality of the speech codec compares well with the codec in the GSM system. The latter operates at 13 Kbps, whereas the CELP codec only requires 8 Kbps. This is partly due to the progress in the speech coding field during the last two years and partly to the greater complexity of the CELP codec compared to the GSM codec.

9.3.2 Modulation

It was decided from the beginning that three users would share one carrier. Two users per carrier would have meant too limited a capacity increase, while four users would have been extremely difficult to squeeze in. With a total bit rate of approximately 16 Kbps per user, including signaling and overhead, the bit rate for one radio channel would be 48 Kbps. Proposals varied between 42 and 54 Kbps. The designers soon decided on 13 Kbps as a working hypothesis for the combined speech and channel coding process. Hardware development and implementation of the proposed speech and channel coder could then start.

The transmission of 48 Kbps in a 30-kHz signal requires a modulation scheme that accommodates 1.6 bits/Hz. This would not have been possible with the previously most commonly used constant envelope modulation (e.g., GMSK, which was chosen for GSM). In the United States, a modulation method with a nonconstant envelope was chosen: $\pi/4$-shifted differentially encoded QPSK, which achieves the desired modulation efficiency. Since the modulation is differential, simple, noncoherent receivers can be used.

The pulse shaping was determined 6 months later when the total bit rate had been selected (48.6 Kbps). The pulse shaping determines the width of the radio spectrum. The U.S. filter choice was raised cosine with a rolloff factor of 0.35.

In order to be able to transmit the three speech channels in a bandwidth of 30 kHz, it is necessary to use this nonconstant envelope modulation method—the first to be used in a cellular system. This method requires a somewhat more complex power amplifier and higher power, thus reducing the time between battery recharges for handheld stations. The solution is considered to provide a good balance between system capacity and demands on the mobile units. In Europe, the channel bandwidth was chosen so that a constant envelope modulation scheme could be used. In North America, this was not feasible because of the requirement for dual mode (both analog and digital speech channels).

9.3.3 Frame and Time Slot Structure for Traffic Channels

The frame structure is shown in Figure 9.3. The structure is also prepared for half-rate coders so that future advances in speech coding technology can be exploited. At the initial stage, when full-rate channels are used, the TDMA frame has a length of 20 ms. The time slot length is 20/3 ms for both full- and half-rate channels. The time slot formats for traffic to and from the base station are different (Figure 9.4). The time slot contains

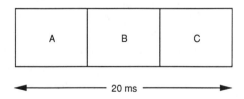

Figure 9.3 The frame structure of the traffic channels. The length of the TDMA frame is 20 ms. The length of the time slot is 20/3 ms for both full- and half-rate channels. A = time slot, user 1, B = time slot, user 2, C = time slot, user 3.

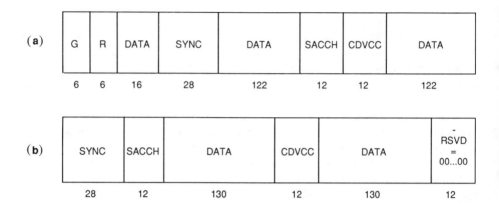

Figure 9.4 The time slot formats are different for traffic to and from the base station: (a) the mobile station is transmitting to the base station; (b) the base station is transmitting to the mobile unit. G = guard time. R = ramp time for the transmitter in the mobile unit. Data = user information or FACCH. SYNC = synchronization and training. SACCH = slot associated control channel. CDVCC = coded digital verification color code. RSVD = reserved.

a 28-bit synchronization word (14 symbols), 12 bits for SACCH—a signaling channel in parallel with the user data—and 12 bits for coded digital verification color code (CDVCC). Digital verification color code (DVCC) is an 8-bit identifier that provides facilities for separating subscribers who use the same physical channel but who are controlled by different, nearby base stations, the so-called cochannels. DVCC is protected by a shortened Hamming code to form the 12-bit CDVCC.

The guard time, corresponding to six symbol periods, is needed to prevent adjoining time slots transmitted by mobile units from colliding into the base stations. Such collisions may occur if the time slots are exposed to variations in propagation time on the way to the base station. The base station can adjust the transmission from the mobile unit in steps of half a symbol period; it sends a time alignment message. The length of the guard time is set to avoid collisions in cells with a radius of up to approximately 10 miles. Thus, there is no need to fine tune the mobile unit transmission during handover between cells of normal size. If the mobile unit is to send a shortened burst as its first signal to the base station, it is told so in the handover command. This first signal has been introduced solely to determine how the transmission time of the mobile unit needs to be adjusted.

The mobile unit does not transmit any data until it has received a time adjustment message. The procedure delays the handover, but it is needed only in large cells. The guard time was chosen as a tradeoff between efficiency (short guard time) and a fast handover procedure (long guard time).

Each connection on a carrier has a unique synchronization word, but the different carriers use the same set. The different synchronization words make it possible for base stations in an area to identify a connection—frequency and time slot. The words are used for verification during handover. CDVCC can be used to distinguish a strongly interfering user (cochannel) from the designated user.

The mobile transmitter has to be turned on and off smoothly to avoid spectrum splatter. The time allocated for power ramp-up is six symbol periods, 123 μs. Guard time is reserved only for the switching-on process. The switching-off time will thus overlap the guard time for the adjoining time slot.

The base station output power has to be kept constant for the duration of the whole frame if a time slot is occupied. The mobile unit can then select the best antenna before its own time slot occurs (preslot antenna diversity).

9.3.4 Associated Control Channels

No digital common control channel has been specified; the standard uses the analog common control channel (see Chapter 2). However, two digital control channels have been defined for user-specific signaling: SACCH and FACCH.

The SACCH contains 12 bits in each time slot. Its overall bit rate is 600 bps and it is used mainly for sending measurements from the mobile station (see Section 9.3.6). The FACCH is used for urgent signals such as handover orders, and when this is to be transmitted, it replaces a speech block.

9.3.5 Channel Coding and Interleaving

A speech block consists of 159 bits divided into three classes: Class 1a (12 bits), Class 1b (65 bits), and Class 2 (82 bits), as shown in Figure 9.5. The Class 1a bits are the most significant and therefore require more protection.

Class 1a bits are protected by a 7-bit CRC code. The protected Class 1a, Class 1b, and 5 tail bits are encoded with a half-rate convolutional code with a constraint length equal to 6. The resulting 178 bits, together with the Class 2 bits, form a block of 260

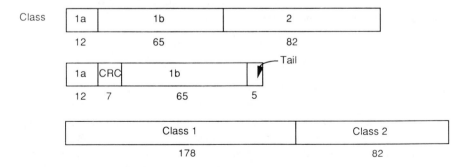

Figure 9.5 Channel encoding.

bits. This block is then interleaved diagonally, and thus self-synchronized, over two time slots. SACCH is also protected with a half-rate convolutional code but is diagonally interleaved with a depth of 12.

The 50-bit SACCH message is protected by 16 CRC bits and then encoded to a total of 132 bits. A coded message is distributed over 22 time slots by the interleaving process. Channel coding and interleaving are continuous processes and do not require block synchronization.

Message synchronization in the decoded bit stream at the receiver is accomplished by checking the CRC code for several positions until it indicates correct received information.

The 49-bit FACCH message is protected by 16 CRC bits and then encoded to a total of 260 bits with a half-rate convolutional code. No tail bits are used in this encoding, the purpose being to facilitate the use of a quarter-rate code for the 49-bit message length. The low encoding rate (i.e., high redundancy) permits signaling—for example, to support handover commands—even with a radio link quality that gives very distorted audio reception.

No explicit flag is used to distinguish speech blocks from FACCH frames. Such a flag would be vulnerable to bit errors or would require a large overhead. The CRC bits used for speech and signaling are used instead, which gives a high degree of certainty.

9.3.6 Mobile-Assisted Handover

This function is mandatory for all mobile units. It is initiated by the base station. The mobile unit measures the received signal strength (RSS) and the bit error rate on its traffic channel. It also measures the RSS on up to 12 other channels. The mean value of the measurements is prepared in the mobile unit and is transmitted in the SACCH once every second. In the case of discontinuous transmission (the transmitter is switched off during speech pauses), the channel quality information is sent via the FACCH. The RSS on other channels is measured in the idle time slot (the spare time in every 20-ms TDMA frame after the message is sent from the mobile unit and before the slot where it receives information). There is a small offset between transmission and reception, which the mobile unit can use for preslot antenna selection diversity.

9.3.7 Time Dispersion

The radio signal can be reflected by hills, buildings, and so on at a distance from the mobile unit, so the received signal usually consists of a direct signal and a delayed echo, as shown in Figure 9.6. Intersymbol interference then occurs if the delay amounts to a major part of the symbol time (41 µs). This results in severe distortion of regenerated data unless action is taken. Antenna diversity can take care of moderate time dispersion, but major delays require equalizers. The greater complexity introduced by equalizers was considered a problem by the manufacturers, but the ability of the system to operate in all

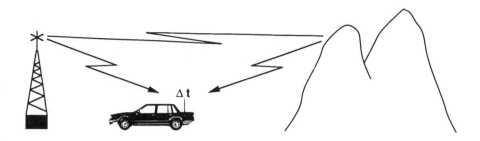

Figure 9.6 Time dispersion.

areas with the radio base stations on the same sites as the analog network stations was greatly appreciated by some network operators. After long discussions, it was decided that the receiver must be able to handle delays of the same length as the symbol period, 41 μs.

Whether this requirement is really necessary or perhaps not stringent enough was debated vigorously. The corresponding value for the GSM system is 16 μs. The North American system is thus less vulnerable to time dispersion.

9.4 FUTURE ENHANCEMENTS

The North American digital cellular system standards are continuously amended. Specifications for authentication and voice privacy have been added to the first version of the standards and work on SMSs, digital fax, and data transmission are scheduled for completion in late 1993. The digital control channel, which will make possible digital-only mobiles, will also be specified at that time. The digital control channel is based on the ISDN call model, which means that future ISDN features can be easily accommodated. For all-digital mobile stations, a new power class of 0.25W is introduced. The battery-saving features of the digital control mobile stations will also allow them to be considerably smaller.

The present digital cellular standard gives a capacity increase of three to four times over that of analog. A number of capacity enhancement techniques are being evaluated that will bring capacity up to ten times. When the technology for half-rate (4 Kbps) speech codecs is available, the resulting capacity increase will be around twenty times.

Chapter 10
Japanese Digital Cellular Radio

N. Nakajima
NTT Mobile Communications Network, Inc.

A shortage of available frequency spectrum due to rapidly growing demand has led to the development of digital systems in Japan as in Europe and North America. In addition, although the mainstream in public cellular telephone systems has until now been voice communication, there is growing demand for a variety of new, high-quality nonvoice services. ISDN digital networks are being installed on fixed networks. Expectations are high for a digital cellular network to provide connectivity with fixed ISDN. Furthermore, the digitalization of the mobile communication makes it easier to ensure the security of communicated information (e.g., by encryption). This chapter presents the emerging Japanese Digital Cellular System.

10.1 INTRODUCTION

There are at present three different analog cellular systems in Japan: the NTT and JTACS systems, and a new NTACS system (for more details, see Chapter 5). This makes it difficult for users to roam between the systems. An agreed-on air-interface specification is needed to permit roaming. The Japanese Ministry of Posts and Telecommunications therefore referred these subjects to the Digital Mobile Telephone System Committee of the Telecommunications Technology Council. In June 1990, the committee issued its report on the technical conditions for a digital cellular system featuring efficient spectrum utilization and flexibility of system functions. In April 1991, the detailed standard for JDC was established by the Research & Development Center for Radio Systems (RCR) [1]. This standard has unified the air-interface of the digital cellular system in Japan. The air-interface specifications and the system architecture are described in this chapter.

10.2 UNIFIED AIR INTERFACE

The major specifications of the unified air interface of the JDC are shown in Table 10.1 in comparison with those of other digital cellular standards in the world.

The development of narrowband modulation and low bit rate codec enable more effective use of available frequency spectrum compared to analog cellular systems. In addition, digital signal processing, such as FEC and ARQ, promise to improve the speech and data communication quality.

The modulation is $\pi/4$-shifted QPSK, which was chosen because of its efficient spectrum utilization and ease of linear power amplification and detection.

TDMA, adopted for digital mobile communications in Europe and North America, was also recommended for use in the JDC system. The major reasons for this are that (a) time division multiplexing offers greater savings in base station equipment cost and space, and that (b) monitoring the receiving level from surrounding base stations in unused TDMA slots by the mobile station enhances call control and thus enables the construction of smaller cells.

The number of TDMA slots is chosen as 3. In the future, this number will be increased to 6 once a half-rate speech codec has been developed.

The traffic channel bit rate (11.2 Kbps) was determined considering both voice quality and adjacent carrier interferences.

Table 10.1
Digital Cellular System Specifications

	JDC	*ADC*	*GSM*
Frequency			
Forward	810–826 MHz 1477–1489 MHz 1501–1513 MHz	869–894 MHz	935–960 MHz
Reverse	940–956 MHz 1429–1441 MHz 1453–1465 MHz	824–849 MHz	890–915 MHz
Carrier spacing	25 kHz interleaving	30 kHz interleaving	200 kHz interleaving
Modulation	$\pi/4$ shifted QPSK	$\pi/4$ shifted QPSK	GMSK
Multiple access			
Full	3 ch TDMA	3 ch TDMA	8 ch TDMA
Half	6 ch TDMA	6 ch TDMA	16 ch TDMA
Carrier bit rate	42 Kbps	48.6 Kbps	270 Kbps
CODEC			
Full	11.2 Kbps (VSELP)	13 Kbps (VSELP)	22.8 Kbps (RPE-LTP)
Half	5.6 Kbps	6.4 Kbps	11.4 Kbps
Equalizer	Optional	——	Mandatory
Diversity	Optional	Optional	——
Frequency hopping	——	——	Optional

The selection of a speech codec was competitive, following the trend in Europe and North America, and several companies both inside and outside Japan proposed their own codecs. The vector sum excited linear prediction (VSELP) coding algorithm was selected as a result of voice quality opinion tests (mean opinion score evaluation) and a comprehensive evaluation of the practicality of hardware and software. Figure 10.1 shows the voice quality versus BER under fading.

Since the frequency bands for the existing analog cellular systems have already been allocated, new bands have to be assigned for digital systems. They are the 800- and 900-MHz bands with 130 MHz of duplex separation and the 1.5-GHz band with 48 MHz of duplex separation, as shown in Figure 10.2.

Because of the plan to swap out the analog channels, carrier by carrier, with those of the digital system, the carrier frequency separation was chosen to be 50 kHz (25 kHz with interleaving).

The analog system developed in Japan [2] (see also Chapter 5) has an extremely high spectral efficiency (6.25-kHz separation interleaving), and an even better frequency utilization is required for the digital system. It is expected that the TDMA will fulfill this requirement with the introduction of a half-rate speech codec and smaller cells through more advanced call control.

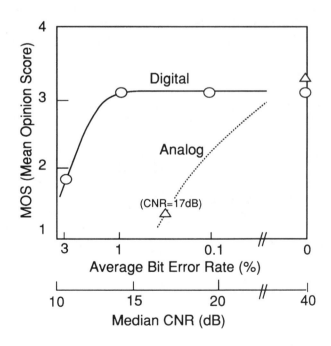

Figure 10.1 Voice quality.

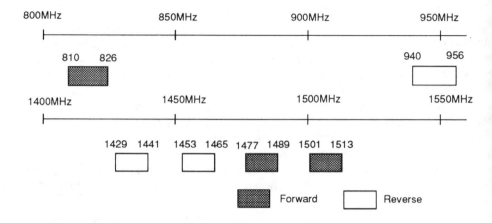

Figure 10.2 Frequency allocation for digital cellular systems.

10.2.1 Logical Channel Structure

A logical channel structure should be designed so that functions can be properly assigned to each channel in order to maintain system flexibility for future service enhancement. Figure 10.3 shows the logical channel structure, which is based on the concept of functional modules defined by CCITT. Details of each control function are given in Section 10.2.4.

10.2.2 Radio Channel Structure

Figure 10.4 shows a TDMA frame structure of a traffic channel applied for communication and a control channel applied for call control. The guard time between the TDMA bursts

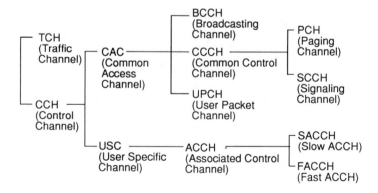

Figure 10.3 Logical channel structure.

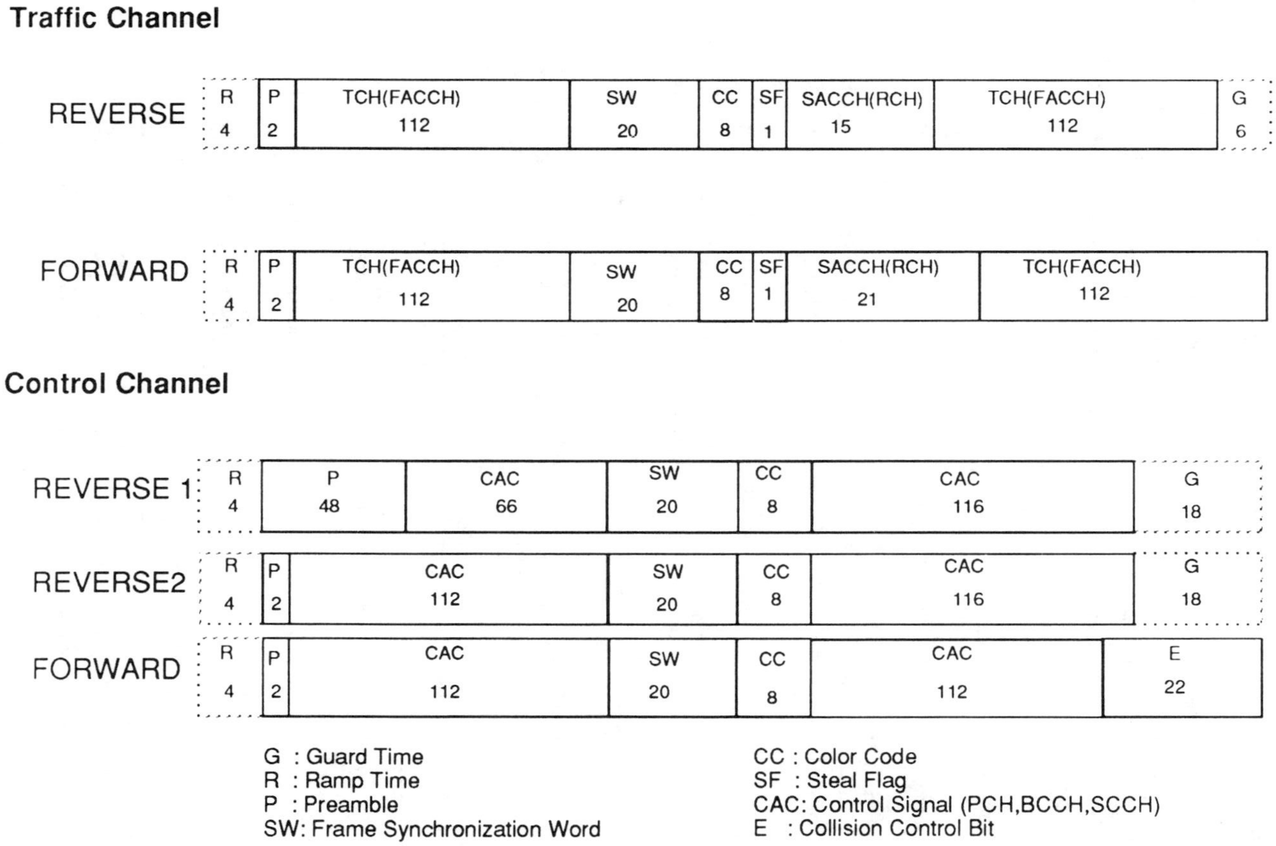

Traffic Channel

REVERSE

| R 4 | P 2 | TCH(FACCH) 112 | SW 20 | CC 8 | SF 1 | SACCH(RCH) 15 | TCH(FACCH) 112 | G 6 |

FORWARD

| R 4 | P 2 | TCH(FACCH) 112 | SW 20 | CC 8 | SF 1 | SACCH(RCH) 21 | TCH(FACCH) 112 |

Control Channel

REVERSE 1

| R 4 | P 48 | CAC 66 | SW 20 | CC 8 | CAC 116 | G 18 |

REVERSE2

| R 4 | P 2 | CAC 112 | SW 20 | CC 8 | CAC 116 | G 18 |

FORWARD

| R 4 | P 2 | CAC 112 | SW 20 | CC 8 | CAC 112 | E 22 |

G : Guard Time
R : Ramp Time
P : Preamble
SW: Frame Synchronization Word

CC : Color Code
SF : Steal Flag
CAC: Control Signal (PCH,BCCH,SCCH)
E : Collision Control Bit

Figure 10.4 TDMA frame structure.

is 6 bits, which corresponds to a maximum cell radius of 20 km. For more than 20 km, time alignment is provided to avoid interference between the bursts coming from mobile stations.

To suppress the interference between bursts caused by the up and down ramp burst amplitudes, an additional 4 bits are assigned. A synchronization word (20 bits), also used as a training word for multipath fading equalization, is placed in the middle of the burst. A color code is used to identify the current base station. A steal flag is used to discriminate FACCH from TCH.

10.2.3 Signaling Structure

The following are required for the signaling.

- Spectrum utilization is efficient.
- A variety of enhanced services such as ISDN can be offered with flexibility.
- The function of each layer should be defined according to the open systems interconnect (OSI) model.
- Layer 3 should be based on the CCITT and CCIR recommendations.

The signaling structure is divided into three layers, as indicated in Figure 10.5. Layer 2 consists of an address part and a control part. The address part is used to allow

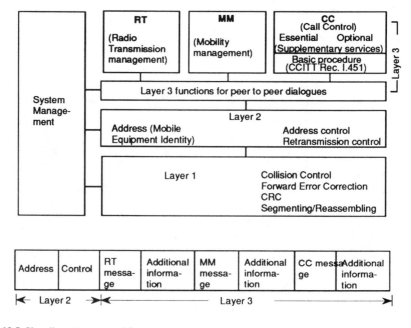

Figure 10.5 Signaling structure and format.

a base station and multiple mobile stations to communicate link control information simultaneously by a common control channel (CCCH). The control part is used for retransmission of the signals. In accordance with the CCITT recommendations, layer 3 is divided into three modules, namely:

- Call control (CC);
- Mobility management (MM);
- Radio transmission management (RT).

To improve transmission efficiency, CC, MM, and RT messages are combined into a single packet (Figure 10.5).

The CC structure is based on the I-interface of layer 3 (CCITT Rec. I. 451) as standardized by CCITT. However, some supplementary service procedures for the I-interface require a large amount of information. The associated control channel (ACCH) signal speed is low (around 1 Kbps), so that these service procedures cause a large transmission delay. To reduce this delay, a shortened procedure that is specialized to mobile radio is provided in addition to the I-interface procedure.

10.2.4 Control Channel Function

As shown in Figure 10.3, control channels consist of a common access channel (CAC) and a user-specific channel (USC). The CAC consists of a BCCH, a CCCH, and a user packet channel (UPCH). The CCCH can be classified into channels for paging (PCH) and for signaling (SCCH). Base stations use individual frequencies for CACs. BCCH, PCH, and SCCH are mapped onto the same physical channels.

- UPCH is applied for packet communication services.
- SACCH carries control information, such as receiving level and interference level, which are used for radio link control during a call.
- FACCH carries control information by interrupting the call.

10.3 SYSTEM CONFIGURATION

It is only the air interface of the digital system that has been standardized in the RCR. The design of the network and terminal equipment is left to the operators or manufacturers. The network configuration and system construction of the NTT DoCoMo system are outlined in this section.

10.3.1 Network Configuration

The digital cellular communications network should fulfill the following requirements.

- Connectivity with the fixed network (analog telephone network and ISDN);

- Roaming of terminals between different cellular communications networks;
- Connection of the mobile and fixed networks by means of a unified interface;

A network configuration as shown in Figure 10.6 is used for the digital cellular systems. The network consists of the mobile communications control center, the base station, and the mobile station. The MCC comprises three parts, which are logically independent of each other. They are a gateway mobile communications control center (G-MCC), providing a toll switch function and gateway to the fixed network; a visit mobile communications control center (V-MCC), providing location management and call connection functions; and an HLR, in which the mobile station numbers and areas where the subscribers belong are registered. A CCITT No. 7 signaling system, which is a common channel signaling system separating the control signal path from the traffic path, is used for control signal transmission between control stations.

When the mobile station moves from its home cellular network to another cellular network (roaming), it is necessary to track the connection between networks and to confirm that the roaming mobile station is an authentic user. To provide this connection tracking

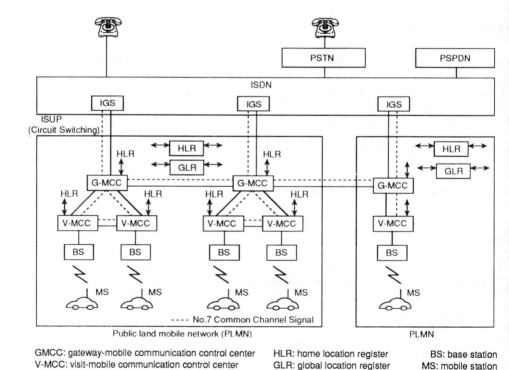

GMCC: gateway-mobile communication control center HLR: home location register BS: base station
V-MCC: visit-mobile communication control center GLR: global location register MS: mobile station
IGS: interworking gate switch

Figure 10.6 Digital cellular network configuration (as used by NTT in Japan).

ability in the NTT DoCoMo system, a gateway location register (GLR) is provided in the HLR to temporarily store the data of a mobile station moving in from another network.

When there is a call from a fixed telephone to this mobile station, the fixed network accesses the mobile station's HLR at first and finds the cellular network, where the mobile station temporarily belongs. Then the fixed network accesses the GLR of that cellular network and finds the V-MCC, where the mobile station is located. Finally, that MCC pages.

10.3.2 ISDN Connection

CCITT SGXI adopted recommendations for digital mobile communications during the period 1985 to 1988 and recommended basic system designs and an interoffice signaling system (Q.1000 series). These recommendations also defined methods for interconnectivity with ISDN in conjunction with the establishment of ISDN in the fixed networks.

As for current analog cellular communication systems, connection with the fixed telephone network is accomplished by using an associated channel signaling system and common channel signaling system. However, in a digital cellular communications system, all connections with the fixed network (ISDN, analog telephone networks) are established through the interworking gate switch (IGS) using No. 7 signaling system ISDN user part (ISUP). This means that, while there are certain limitations, the wide range of ISDN services can also be provided over a cellular communications system. In addition, by using a standard international interface, cellular communications networks can be easily interconnected.

10.3.3 Diversity Reception

Diversity improves the transmission quality against interference. Figure 10.7(a) shows that 2×10^{-2} BER is obtained under CIR = 11 dB. If diversity were not employed, CIR would need to be 17 dB.

Figure 10.7(b) shows that diversity is also effective for multipath fading. Measured delay spread distributions in various areas are shown in Figure 10.8. In the case of mountainous areas, experiments have shown that a delay spread of over 5 μs results in a BER that exceeds 2×10^{-2} for 10% of the area if diversity is not used. With diversity, the $>2 \times 10^{-2}$ BER probability drops to less than 2% of the area. These results suggest that an adaptive equalizer may not be necessary if diversity is employed.

10.3.4 System Configuration

The system configuration of the NTT DoCoMo digital cellular communications system is shown in Figure 10.9. This system comprises the MCCs, the base stations, the mobile stations, and the mobile communications operations centers (MOC).

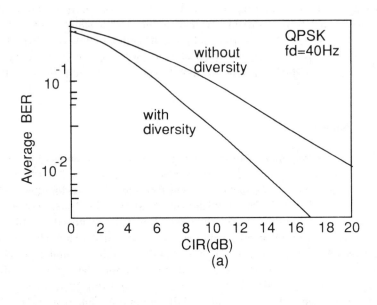

(a)

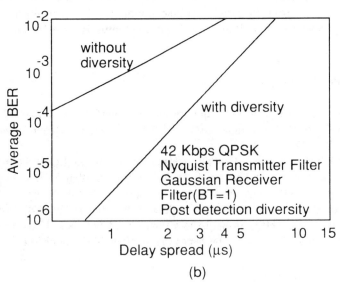

(b)

Figure 10.7 Diversity effect: (a) average BER vs. CIR; (b) BER vs. delay spread.

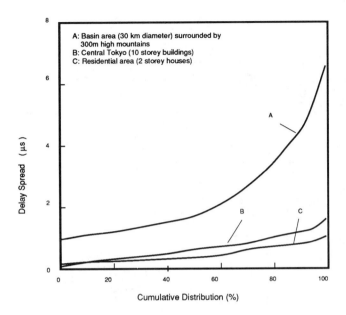

Figure 10.8 Delay spread measured at 1.5 GHz.

The MCC comprises switching system, the digital automobile switch (D-AMS), speech processing equipment (SPE) for speech coding and decoding for radio links, and the mobile controller equipment (MCE) for paging and other signal distribution. Since the SPE performs low-bit-rate coding and decoding (11.2 Kbps), one 64-Kbps link between base station and MCC can carry multiple traffic channels. In the NTT DoCoMo system, three traffic channels with timing signals are carried by one 64-Kbps link. Thus, economical communications links between MCC and the base station can be provided.

The base station comprises the base station control equipment (BCE), base station modulation and demodulation equipment (MDE), base station transmitter-amplifier equipment (AMP), base station antenna (ANT), and the test transmitter/receiver (TTR). The BCE also can be installed in the MCC. This configuration is effective when it is difficult to ensure sufficient installation space in the base station. In addition, by using a three-channel TDMA in the MDE, the number of components per carrier is one-third that of FDMA. By using the multiple carrier common amplifier, the base station becomes both smaller and less expensive than analog ones.

10.4 SYSTEM FEATURES

10.4.1 Increased System Capacity

Spectral efficiency is estimated by comparing the digital system capacity with that of existing analog systems.

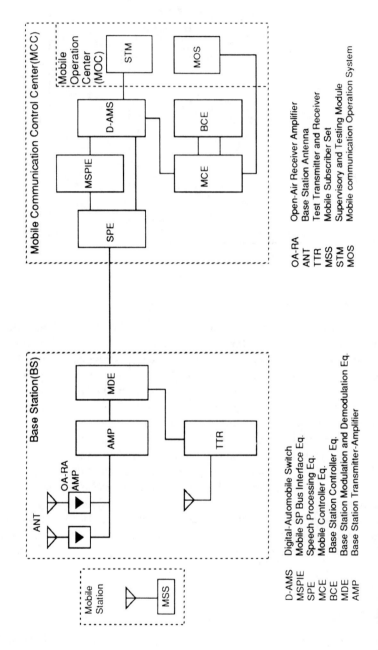

Figure 10.9 System configuration.

OA-RA   Open-Air Receiver Amplifier
ANT   Base Station Antenna
TTR   Test Transmitter and Receiver
MSS   Mobile Subscriber Set
STM   Supervisory and Testing Module
MOS   Mobile communication Operation System

D-AMS   Digital-Automobile Switch
MSPIE   Mobile SP Bus Interface Eq.
SPE   Speech Processing Eq.
MCE   Mobile Controller Eq.
BCE   Base Station Controller Eq.
MDE   Base Station Modulation and Demodulation Eq.
AMP   Base Station Transmitter-Amplifier

In general, the frequency utilization factor is given as the product of the usage rate of the frequency factor and that of the spatial factor. The per-channel frequency bandwidth in the digital system is 8.3 kHz (= 25 kHz/3 interleaved channels), and 6.25 kHz (interleaved channels) in the high-capacity analog system [2]. Thus, the usage rate of the frequency factor of a digital system is 0.75 (= 6.25/8.3) compared to that of an existing high-capacity analog system. The frequency factor of the digital system is thus slightly less than that of the analog systems.

The frequency reuse cell number, which is a measure of spectrum utilization in the spatial dimension, is a 7-base station (site) with 3 sectors with current analog systems. The digital system, however, provides robustness against cochannel and adjacent-channel interferences by means of codec and error correction technology. As a result, 4-site by 3-sector cell reuse is possible. The usage rate of the spatial factor is improved 1.7 times (= 7/4) over a conventional high-capacity analog system.

Consequently, the spectrum utilization of the digital system will be around 1.3 times (= 6.25/8.3 × 7/4) that of the conventional analog system, and system capacity can thus be increased when compared to current analog systems.

In addition, future half-rate codec formats, flexible reuse of frequencies, and smaller cell configuration can be expected to yield further improvements in spectrum utilization with digital systems.

10.4.2 Improved Communications Quality

While speech quality gradually deteriorates as the signal reception level drops in an analog system, it should be possible to maintain consistent quality up to a certain error rate with a digital system (Figure 10.1). Although the error-free (a high receiving level) quality is slightly worse than that of analog systems, digital systems can be expected to provide better speech quality than analog systems over the total service area.

Digitalization is also effective when it comes to security and confidentiality of communications. While eavesdropping is difficult with voice coding/encoding alone, the combined use of signal scrambling and encryption offers an extremely high level of security and confidentiality. Furthermore, a major feature of digital communications is that voice encryption introduces no deterioration of speech quality.

Error-free communication is achieved in facsimile transmissions by adopting the high-throughput ARQ transmission scheme [3]. A signal processing unit is provided on the mobile stations (optional) and switching equipment for error correction for 4.8-Kbps transmission through an 11.2-Kbps channel.

A signal processing unit for the widely used micronetworking protocol (MNP) modem protocol is provided for data communications. Since a retransmit function is already provided in the modem, only forward error correction is applied in order to improve transmission quality and throughput.

10.4.3 Smaller and More Economical

It is expected that the base station equipment will be approximately one-fifth the size and significantly less expensive than analog equipment due to the use of a three-channel TDMA format, multiple carrier common amplifier [4].

The mobile equipment will become smaller through the development of LSI and VLSI devices, reducing power consumption with a voice-activated transmitting power switch (VOX), and by adopting diversity reception rather than a time dispersion equalizer to overcome multipath fading.

10.4.4 Future Services

At the introduction of this system, speech and data (modem) communications will be the primary concern, but new services capitalizing on the features of digital technology will gradually be introduced, such as short messages, ISDN unrestricted digital information, radio packet communication services, and others.

REFERENCES

[1] "Digital Cellular Telecommunication System RCR Standard," Research & Development Centre for Radio Systems, STD-27, 30 April 1991.

[2] Kuramoto, M., K. Watanabe, M. Eguchi, S. Yuki, and K. Ogawa, "High Capacity Land Mobile Communication Systems in NTT," *Journal of IEICE Japan*, Vol. 71, No. 10, October 1988, pp. 1011–1022.

[3] Ito, S., K. Sawai, S. Uebayashi, and T. Matsumoto,"G3 Facsimile Signal Transmission Using WORM-ARQ in Digital Mobile Radio," *IEICE Technical Report*, CS91-26, 1991.

[4] Narahashi Y., and T. Nojima, "Extremely Low-Distortion Multi-Carrier Amplifier (SAFF)," ICC'91, June 1991, pp. 1485–1490.

Chapter 11
Cellular Radio Planning Tools

A. Bajwa
Mobile Systems International

11.1 INTRODUCTION

Competitive mobile services, of which cellular is an important market, have begun to come under considerable commercial pressure from consumer-oriented new entrants like PCNs and personal communications services (PCS). Regardless of the radio access technology, the cellular component of these systems still represent the dominant infrastructure outlay costs. Startup systems must provide wide-area coverage. Network operators are now acutely aware of the cell count and the cost of equipment provisioning. Operational networks are continually striving to maintain call quality against increased capacity demand. This means optimizing and maintaining a complex network, while always being flexible to customer demands, which remain largely unpredictable in usage as the market matures and the deregulated industry tries innovative marketing ploys. In response, cellular engineering must be technologically creative to present solutions to handle this complex task in an efficient manner. Cellular planning tools are increasingly emerging as important weapons to ensure cost-effective investments and near-optimized network performance. Advanced interactive tools are now available for guiding the planning and operations engineers and managers in designing high-quality networks to deliver high capacity to end users of mobile services.

In this chapter, we illustrate how the use of advanced software tools efficiently translates the customer's requirements into a viable cellular radio network. The text and accompanying illustrations take the reader through the steps in a generic process of cellular planning modeled on a startup network. Some basics are introduced in the form of definitions, but the reader is referred to Chapter 1 for further information on the techniques of cellular engineering.

11.2 BASIC DEFINITIONS

Communications with customers on the move inherently makes the definition of service coverage probabilistic in the sense of both radio and traffic characteristics. A network offers service in a geographical area to a customer who chooses to be anywhere in this *coverage* and can make or receive a call. This calls for a definition of coverage in relation to a quality of service (QOS) criterion, and simultaneously for a measure of the system capacity such as the grade of service (GOS). The thresholds for these customer requirements are determined from the marketing strategy of the network operator and, in competitive markets, by certain license obligations set by the telecommunications regulator. The choice of QOS and GOS have a profound impact on the network costs and the network revenue.

11.2.1 Quality of Service

Call quality is linked to the probability that the customer can initiate, receive, and maintain an acceptable call in the service area, also known as coverage. It is well known that radio signals fluctuate due to fading or multipath propagation. The signal also naturally attenuates with distances from the base station. The position of the mobile customer is random, and therefore the nature of the environment affecting signal propagation is changeable. Furthermore, there is interference due to unwanted signals from other customers who are reusing the same (cochannel) or adjacent channels in other cells in the network. The interference levels are also related to the traffic in the network, and it becomes immediately apparent that the analysis is extremely complex.

Traditionally, the QOS is defined as the percentage of acceptable calls in a percentage of locations in the service area. To establish the threshold QOS, the power link budgets of the downlink and uplink paths from the base station are calculated. Those which exceed the threshold required for an acceptable call are used to determine the area coverage (i.e., percentage of locations criterion). The link budget calculations are determined by the mobile power class, since in almost all cases the uplink from the mobile to the base station is the weaker link. The coverage is therefore always referenced to the mobile power class.

11.2.2 Grade of Service

The system GOS is usually defined as the busy-hour call blocking probability. It is an indication of the ability of the system to meet the traffic demand. Clearly, if the demand exceeds the provisioned capacity, then calls will be blocked. The GOS in the fixed network part is usually specified to be much higher than the radio access link to ensure that blocking in the fixed part is not significant by comparison. This is done to ensure that the radio channels are efficiently utilized, since spectrum is the scarce resource in a mobile network.

11.2.3 Call Dropout

In any radio system there is always a small probability that a call in progress may be terminated prematurely. The cause for this is not necessarily the radio conditions, although from a system point of view it may be necessary sometimes to terminate a call that has ventured beyond the system domain for coverage or interference reasons (for example, because of an unsuccessful handover attempt by a lower mobile power class than is permissible in the coverage area of the cell).

11.3 CELLULAR PLANNING STEPS

The process of cellular planning for a startup network is different from that for an operational network in which coverage is extended through network rollout or capacity is enhanced by way of cell division. To illustrate the startup network case, it is adequate to demonstrate the development of cellular planning with the aid of the software tools.

The first step in the process is to analyze the customer requirements. The assumption is that the network operator has marketing forecasts on the number of customers, the customer growth, and the customer usage. The business plan then iteratively translates this to a rollout of the network to meet the investment goal. This is followed by an initial cell layout, also sometimes known as the *nominal cell plan*. In some networks, the transmission network between the cell site and the BSC, and between the BSC and the MSC, is provisioned with private millimetric or microwave radio links. If this is a requirement, then a microwave planning tool is also used to analyze the path and determine the equipment configuration.

The nominal cell plan is used to find cell sites, and thereafter the cellular planning process is interlaced with the property acquisition process. The site candidates found to be consistent with the search areas in the nominal cell plan are then evaluated from a radio acceptance point of view. This evaluation is conducted partly using the planning tool to find the best candidate. Candidate sites are also evaluated to meet the installation requirements (e.g., the civil works and provisioning of power, transmission plant, etc.). Site acquisition then commences, and the process continues until all sites are selected. Frequency planning can start once the sites have been selected. It is not until the final commissioning is due that the initial handover and power control cell parameters' settings are determined prior to the system integration tests.

It is not possible to go through the complete interactive cycle of this process in this chapter, but it is practical to illustrate the outputs from the planning tool for the important stages in development of a cell plan.

11.3.1 Analysis of Customer Requirements

The marketing plan is the usual starting point. It contains the customer projections on an annual basis, typically showing year-end figures on the number of customers and the usage profile per customer to be used. Additionally, a geographical coverage is shown together with the QOS requirement for each mobile power class in the network. The GOS is also specified, and sometimes additional geographical differentiation of customer traffic projections is also available. Clearly, the more reliable the information the more realistic the cell plan is. Most often the network operator has a network rollout plan determined by the service marketing and the business plans. The QOS requirements for each mobile power class are used to calculate the link power budget. This is then converted to minimal cell radii for the initial assessment with a simple spreadsheet model for the cell count. Most tools can provide a facility through the use of a menu to investigate the uplink and downlink power budgets. System gains and losses are calculated from the values entered for each of the link parameters (e.g., feeder loss and combiner loss). The transmit and receive power levels are calculated based on the figures supplied by the user and the maximum allowed path loss derived for the QOS threshold.

11.3.2 Initial Cell Layout

The creation of an initial layout of sites for cells is a primary function of the planning tool. All tools overlay cells on a background of a digitized map. The basic feature is the digital terrain map (DTM) with height data on which can be superimposed vector information such as roads and coastline, and clutter data derived from land usage. The user interfaces for these tools are intuitive with a pop-up menu facility. The user of the tool can decide the depth and detail of the display; for example, to show all clutter categories available in the database or only those relevant to the coverage calculations. To start the process of cell layout, a facility known as *hexagons* is available; the user can add a cell at a specific location with a notional hexagon of a prescribed radius, move such a cell about on the screen, or perhaps delete it. The hexagons can be snapped onto a regular grid automatically or moved manually if adjustments to the location are required. The tool allows the user to zoom onto a small area to examine local characteristics, or zoom out to get an overview of the planned area. Associated with each cell is a site database that can also be viewed. This stores various site and cell information, including the propagation model used in the predictions, the antenna characteristics, and also the site status (e.g., phase in the rollout).

The DTM may be set on a universal transverse Mercator (UTM) grid system, with the location of the cell in eastings and northings referenced to each UTM zone. Advanced tools will offer a range of map projections on a grid system other than UTM, but in each case the map has to be accordingly processed. In some cases, it is even convenient to use latitude and longitude (Lat-Long).

In creating the cell layout, the user may use the terrain background database to accurately place sites. Alternatively, the exact site coordinates may be entered, allowing the tool to place the site on the display. In either case, the site database will show the site coordinates in eastings and northings.

Vector information proves equally useful in positioning cells in the initial layout. The vector and clutter classification information can help in the rough selection of areas where greater benefits may accrue in capturing traffic at the initial stage. Later, the user can obtain precise traffic analysis against the chosen layout if the tool supports the facility for traffic analysis. Figure 11.1 shows an onscreen display of the mapping and clutter data.* The map information displayed may be presented as a backdrop for overlays of signal predictions and other modelled parameters. The height data are shown as a block fill on a gray scale. This presentation makes the display of colors effective for the vector and clutter data. Cell locations are shown with a unique cell ID. Note the information picked up by the position of the movable cursor in the bottom right-hand margin of the display.

11.3.3 Coverage Assessment

Crucial to the coverage assessment is the use of an accurate propagation model. To derive the best model, it is usual to process measurements of signal strength to test the accuracy of the mathematical model. An advanced tool will offer a model editor with user-selectable and adjustable parameters. In addition, it will perform regression analyses to verify the best fit to measured radio survey data. Some tools support an integrated facility to enter the measurements directly from disks produced in field radio surveys. The provision of such interfaces and analysis features gives the user freedom to create and optimize the propagation model most appropriate to the radio network.

The mathematical equation typically relates the propagation path loss to the distance from a fixed base station to a service area square (or bin) where a mobile may be located. Path loss due to distance is expressed as a slope and intercept equation with various adjustment factors. The model uses the DTM terrain heights and clutter classifications through a user-adjustable weighting for, say, the clutter classes. In addition, a flexible tool will allow the antenna characteristics to be entered by the user and will support an antenna editor that can accept the three-dimensional antenna radiation patterns.

The signal strength prediction calculation may be done for a single selected cell or a group of cells, with the parameters set in the network database in accordance with the network rollout or phase if applicable. The resolution of the square or bin for each calculation is determined by the DTM resolution, but the square area, over which the prediction is to be performed, is determined by the user according to the need of the network. Signal values for each bin are stored in an array for each cell, and the overall coverage in a large geographical area is created by overlaying the signals from all the cells and determining an overall array. The coverage may be then displayed as an attractive color contour according to the QOS requirement for each mobile power class. This is used to assess coverage, and inadequacies are remedied by, for example, repositioning

---

*All figures in this chapter appear on pages 259–268.

cells. See Figure 11.2. Classes of mobile have been selected here from the GSM 900 standard and color-coded as shown in the accompanying key.

The predicted signal strength array is also used inside the tool to derive the interference levels and, with the overlay of a frequency plan, it provides the cochannel signal levels. Calculations are repeated only when changes are made to the cell parameters of a particular site (e.g., antenna height above local ground).

The choice of the propagation model depends on the model fit against typical measured data. The coverage assessment uses the appropriate model for each category of the environment in the coverage area. Any advanced tool ought to interface to a radio survey facility to load, and interactively analyze, measured data to help the user create the propagation model. Figure 11.3 shows a display of the measurement analysis and modeling in a typical advanced tool. The display shows a measurement route with an overlay of signal strength, color-coded in decibels above 1 mW for each bin. In addition, the display shows the residual error between the measured signal strength and the predicted value, depicted here as triangles and squares of different colors. The residual error is an indication of the propagation model used. Therefore, the analysis is also shown in an adjacent window and can iteratively continue until the rms error is acceptable. The model shown here is for illustration purposes, and much of the skill of the radio planner is needed to intelligently perform regression analyses on measured data to derive good models. It is possible to display and analyze multiple radio surveys from one site. The measurements are accurately positioned on the map and displayed for each road route, and the measured signal in bins is compatible with the prediction, and therefore residual or differences between measured and predicted values can be shown directly. Together with this is a full analysis report, which provides the statistical error estimates for each classification of clutter, to assist in the modeling work.

11.3.4 Capacity Assessment

Each cell in the plan is expected to capture only traffic in its own coverage area. Therefore, the cell boundary will determine the traffic level. In the process of planning, the vital input is the customer usage and the geographical calling patterns inside the coverage area. These inputs are derived from the network marketing plans and used as the input data together with the GOS requirement. The tool must intelligently spread the traffic over the coverage area and then capture the traffic inside the boundary of each individual cell. Only advanced tools provide this capability.

One such tool can automatically analyze the customer traffic forecasts and show the traffic offered by each cell on a map display, and also report, on a cell-by-cell basis, for each site. This makes the dimensioning relatively easy as the number of channels are calculated for each phase of the network. Therefore, the capacity provisioning for a number of growth scenarios is readily analyzed. A typical display of the traffic is shown in Figure 11.4. Contours of the traffic intensity are drawn as vector outlines. The tool can spread the traffic according to certain rules, and this is shown color-coded in the plot. Analysis of the traffic is shown in another menu in the display. This shows the number of actual channels assigned and the call blocking on a cell-by-cell basis. Calculations of the number of channels required to achieve the desired GOS are shown in parentheses. This dimensioning is critical to ensure that the equipment is provisioned with the right number of voice channels. The traffic figures for the cells serve as the detailed input for the design of the transmission network, dimensioning, and the planning of the switching network.

11.3.5 Interference Assessment

Choosing the frequency reuse strategy and producing frequency plans in a high-capacity cellular network is not straightforward. Frequency reuse can cause cochannel interference with undesirable consequences. Regular hexagonal reuse is not realizable in practice, but most plans are arranged on an approximate cell grid to maintain a reuse structure and minimize the possibility of cochannel interference. Cochannel interference analysis in a planning tool environment is performed by recreating an array of cochannel interference from the signal predictions in the coverage area of interest. The initial frequency plan sets the cochannel interferers for each wanted cell. The cell parameters are then adjusted to minimize the overall interference in all the cells in the coverage area. The choice of the frequency plan depends on the expected capacity and capacity growth and seeks to provide the target QOS. The tool can test frequency plan scenarios rapidly and offers a visual interactive analysis of the severity of cochannel interference for each of the frequency or carrier assignments. The site database is supplemented by a carrier database that stores choices of the frequency groups. Each site has a database of carriers associated with each antenna. This allows the sectorized concept of sites in the form of cells illuminated by different antennas. The characteristics of the antenna are defined for each cell in the site database as elevation and azimuth of the chosen antenna model.

A very useful feature, which enables a study of the frequency plan, is the antenna editor. This lets the user accurately model the characteristics of the antenna radiation patterns and identify each by antenna type. Measured data may be entered in some tools. The tool should also model the effects of antenna tilting, which is very simple and does not require a fresh signal prediction for each change in either the antenna model or tilt. The analysis is then relatively quick and allows many parameter settings to be investigated before settling on a preferred solution.

A particular scenario for a cell layout is shown here to illustrate a frequency plan. The cochannel frequency groups are displayed showing the cell coverage areas color-coded according to the carrier group allocated in Figure 11.5. System interference is calculated by assessing the worst-case interferer for each grid square, from all the interfering cells, in the coverage area of the wanted cell. The extent of the cochannel interference may be studied by analyzing the severity of the interference from color-coded levels over the entire coverage area. The cochannel interference array is shown for initial values of the transmitter power and antenna tilt. This shows some areas of severe interference in cell number 3 of site number 3 with CIR lower than 15 dB. The next step in the process is to adjust the antenna tilts and powers to progressively reduce interference. The result of the interactive adjustments is displayed in Figure 11.6, where the improvement in the CIR over the coverage area can be clearly seen (in the lower display area).

Automatic frequency planning is one feature that makes a planning tool very powerful. In the context of a tool, it not only requires fast computation and manipulation capability, but also an "intelligent" algorithm. The decision making process is very complex in practice, and hence human interaction is essential to achieve a viable frequency plan. Automatic or semiautomatic frequency planning may be of practical use for generating candidate frequency plans more efficiently, with inbuilt checks exploiting the powerful database. Various sophisticated algorithms are offered in some tools, but the degree of success suggests that such ideas are not yet mature enough to adopt with confidence.

11.4 ENHANCED CAPABILITIES

An advanced cellular planning tool should be integrated in such a way as to offer the user an enhanced modular capability. The tool must evolve in an integrated software environment, without any changes to the user interfaces, so as to offer new features or enhanced capability of the existing features. Certain capabilities of practical interest are now mentioned to show how the cellular planning tools are developing.

11.4.1 Handover Analysis

Handover analysis is complex, even in analog cellular systems. Digital cellular and PCN systems with GSM 900 and DCS 1800 standards add a further dimension to this complexity. The simplistic approach is to determine the domain of a cell by displaying the equal signal power boundary for each cell. This, however, does not reflect the system dynamics. Some planning tools have dealt with this by analyzing parameters over "typical" short segments of measurements to derive the threshold levels for selected parameters. Others extend this by a simulation approach, with predicted values for selected parameters along a route section. These approaches are limited in practical value in that the handover from a system viewpoint is a statistical event, which must be analyzed over the coverage area, not just route segments, to become meaningful. To do this, the tool could perform a statistical analysis of the most likely handover candidates, with user-definable hysteresis levels for the parameters. This mimics more closely the system behavior, with the percentage probability of handover serving as the system parameter of interest and the hysteresis levels acting as the variables to be adjusted (i.e., the threshold). Display of the probability of handover contours in Figure 11.7 allows the cell boundary for a set parameter offset (i.e., margin).

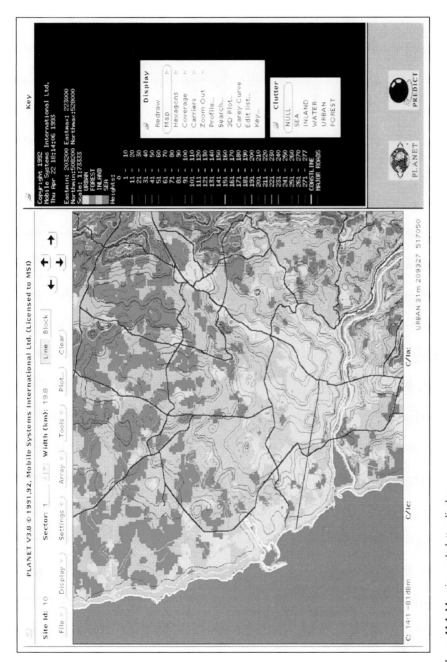

Figure 11.1 Mapping and clutter display.

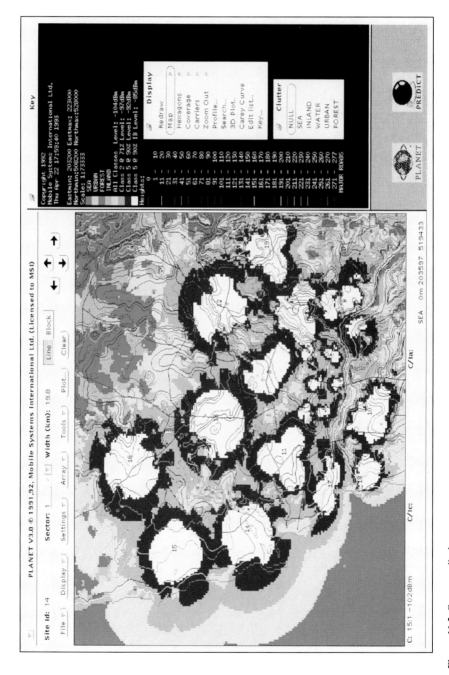

Figure 11.2 Coverage display.

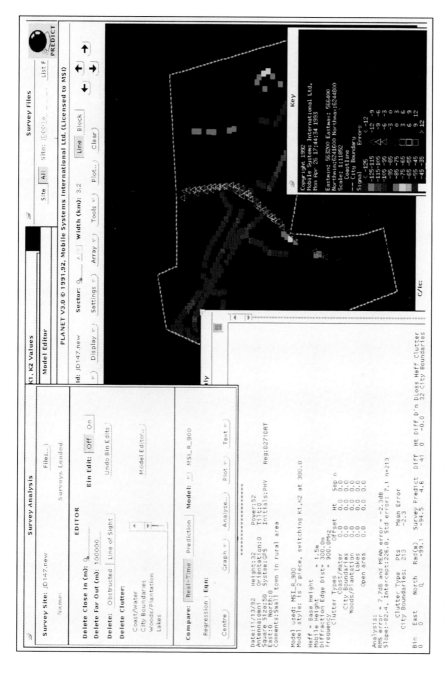

Figure 11.3 Measurement analysis and modeling.

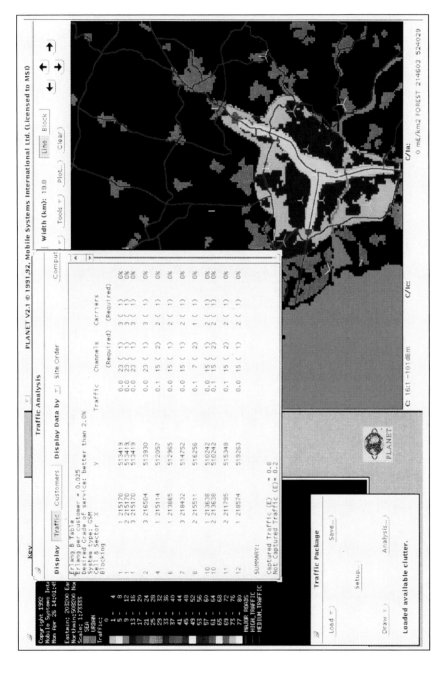

Figure 11.4 Traffic capacity display.

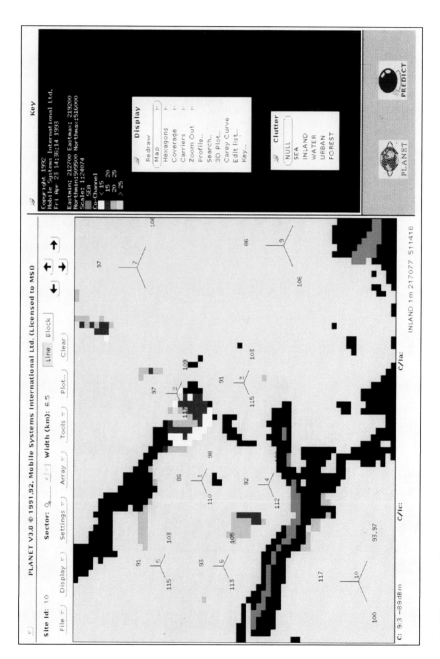

Figure 11.5 Frequency allocation.

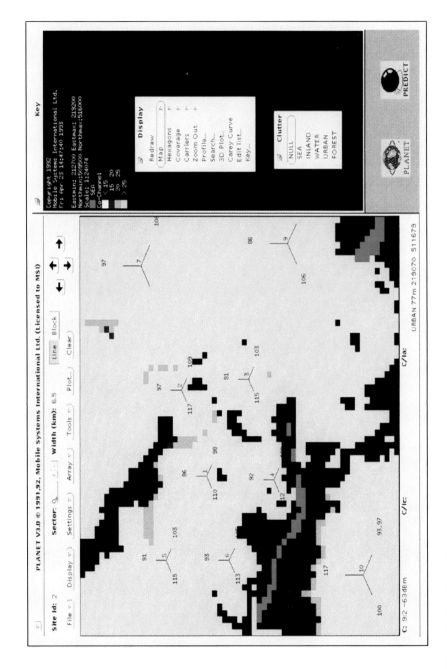

Figure 11.6 Interference improvement.

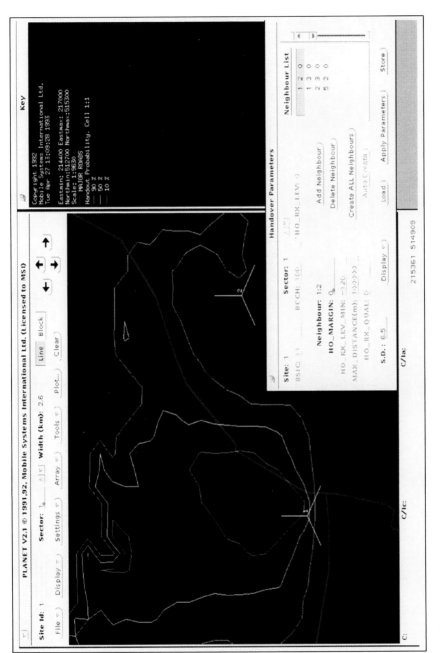

Figure 11.7 Handover contours.

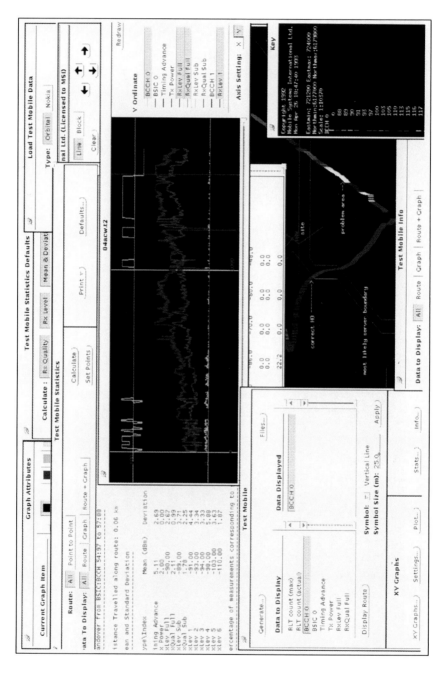

Figure 11.8 System parameters.

267

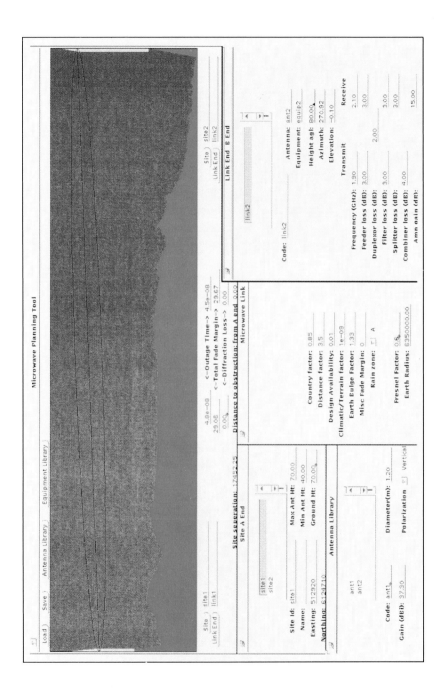

Figure 11.9 Microwave planning.

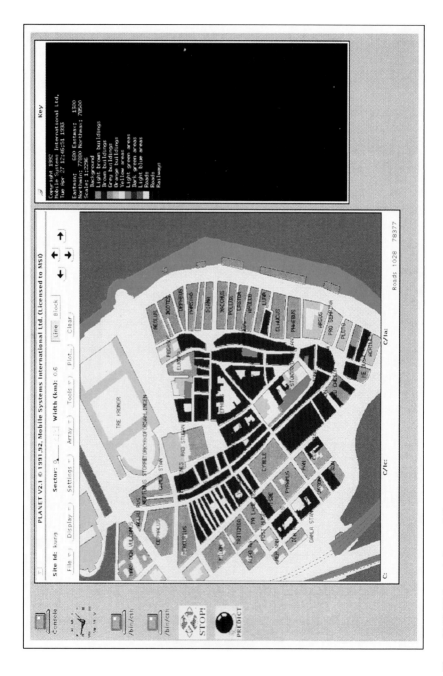

Figure 11.10 High-resolution modeling.

11.4.2 Test Mobile

Test mobiles are essential for testing and optimizing the radio network. Test mobiles are mobile stations with the capability to operate in engineering mode to provide the system signaling and data messages. System parameters like RXLEV and RXQUAL, as measured and reported to the system, can also be monitored. Parameters like this, together with the signaling messages, offer vital clues to the system functionality and performance of the cellular radio network. Therefore, the tool must provide an interface to the test mobile and offer processing and display capability inside the tool environment. Different tools offer different levels of analysis.

Providing statistical analysis capability for the system parameters is straightforward. An x-y plot of the system parameters as a function of distance is shown in Figure 11.8. The parameters can be selected interactively from a menu, and analysis thresholds set as needed for each individually. The display shown here has x-y plots of BCCH0, RXLEV, and RXQUAL. These data can be overlaid on the digitized map and along the test route, where the data was collected by using the Global Positioning System (GPS) positional information. This gives a display that relates to the location of the measured data or event (e.g., handover) in the test route.

11.4.3 Microwave Planning

Microwave planning is closely associated with cellular planning. The provision of micro-wave or millimetric links has to be evaluated for percentage link availability; that is, the maximum hop distance in relation to an assessment of the path profile to meet specified performance limits. The tool must perform automatic link budget calculations for which the user enters the equipment and antenna type, preferably selected from a library. The user can also enter the link parameters, such as frequency, site location, site height, etc., and link performance limits, such as fade margin, or just the design availability figure. A display produced by the microwave planning segment is shown in Figure 11.9. The display shows a two-dimensional path profile between two sites. The first Fresnel zone is marked and the terrain profile is shaded to highlight the undulations in the path. Each site and link end is identified and the site data stored in the database. The display windows show the site information and the antenna and equipment library. Climatic characteristics and other physical parameters that affect the radio path can also be input into the tool.

11.4.4 Microcellular Planning

Microcellular planning is emerging as one of the major challenges in the engineering of high-capacity cellular and PCN systems. To lay the foundation of the propagation and systems models, it is essential to provide the necessary level of map backdrop and database sophistication. The power of the tool is judged by how effectively these capabilities are realized, and not by the characterization of a specific microcell model editor, which gives the user the freedom to enter the specific values determined from an analysis of measured data. Various models of propagation have been studied. Some studies suggest that a Walfish-Ikegami [1] model is appropriate for certain cell layouts. This type of model is easily incorporated into the tool. Microcell technology dictates implementation scenarios that are largely customized to the capability of the system standard. The tool must therefore offer a scenario test-bed to facilitate the widest possible analysis. High resolution maps are essential for this with the various detailed vector and clutter information, forming the basis for models. The capability of the required map handling is demonstrated in Figure 11.10 to show to what extent an advanced tool can easily progress and still remain consistent with the tool concepts. The display shows a small area on a high-resolution map in which buildings and streets are easily identified. Each type of building is shown color-coded, depending on its type, and this information is later used in the microcell propagation model.

11.5 OPERATIONAL SUCCESS WITH PLANNING TOOLS

The performance of a planned cellular network depends on the accuracy of the marketing forecasts, the ability to translate this into effective cellular plans, and success in the implementation rollout. We consider the success of the tool in capturing all the planning requirements in the operational system.

11.5.1 Implementation of Cell Plans

Cell plans are the starting point for the engineering process. The process cycle was briefly described in Section 11.2. The success in the actual implementation of the cell plan is measured by the service quality delivered to the customer in terms of the call success rate inside the coverage area (i.e., the ability to make and receive successful calls within the area served by the cells). To verify the actual coverage and the service quality, it is usual to collect performance parameters and produce appropriate statistics. One way of evaluating system performance is to set up and monitor calls by capturing data from the serving cell and the cell neighbors with a test mobile. The data capture from a test mobile and the import and display capability is described in Section 11.4.2. Subsequent analysis of this data in the planning tool will provide the performance statistics. Alternatively, the performance reports from the O&M system may be used.

The cost of abortive work to remedy problems not only affects service but also has financial implications. Cell plans that are thoroughly and accurately analyzed for practical scenarios are expected to achieve project and network objectives with lower risks. A practical cell plan is one that is flexible enough to adapt readily to unexpected capacity growth or to modified coverage rollout plans. A good tool offers the means to accurately model and test all the practical scenarios in a quantitative manner.

11.5.2 Operational Cell Count

On the basis of outdoor on-street coverage, cell plans produced by experienced cellular engineers using calibrated planning tools ought to achieve accuracy better than 95% against, for example, a target QOS for 90% successful calls in 90% of the coverage area. Practical implementations tend to have fewer cells than those estimated from simplistic spreadsheet calculations. To quantify this with a cell count error is complex, and in practical situations it will vary with the mobile power class and the type of environment in the coverage area. Getting accurate coverage depends on accurate propagation and traffic models and the cell dimensions. Experience has shown that planning with tools will always give consistently good results, since there is scope to go through many "what if" scenarios. Radio planners with moderate skills can produce viable cell plans with a good tool. This would not be possible otherwise, and explains why there is a great demand for good tools. Frequency planning is particularly suited to a tool where complex interactions between cells need to be analyzed.

Cellular radio planning is only the first step in the life cycle of the network. Therefore, it is not always possible to give global indications for the generalized case. Tools are currently being developed to effectively deal with the operational role of optimization and performance engineering.

11.6 CONCLUSIONS

Advanced software tools are essential for planning cellular radio networks, particularly where competitive pressures place a high premium on cost-effective implementation. Base station equipment costs are the largest element of the investment and coverage quality is key to the success of new competing networks.

A sophisticated software suite that brings together all elements of radio planning in a user-friendly manner to the engineer or manager is an industry requirement. Moreover, presenting an integrated platform with relational databases from engineering data to digitized mapping or other morphological data is part of this challenge. Tools now exist that fulfill this need and are rapidly being developed with new and efficient features.

Some important capabilities have been described with the aid of illustrations in this chapter to show the relative ease with which a cell plan can be produced. However, it is not possible to demonstrate in the text the flexibility of use of an interactive environment. The main emphasis here is to show with onscreen displays that such an environment exists in advanced tools.

Tools are now progressing from the domain of planning into operational use. The GSM cellular standard has been used to illustrate the role of a test mobile and the way in which such tools will progress towards the goal of performance engineering. Future standards will also be complex whether they are based on FDMA, TDMA, or CDMA radio access technology. The concept of the tool will be viable in all cases, although the emphasis may shift; for example, frequency planning in microcellular CDMA will be of a different complexity to that in macrocellular FDMA. Future tools must be developed that offer the planning or operational user of the tool the capability for all present and future systems in an integrated facility in an efficient way.

REFERENCE

[1] "Urban Transmission Loss Models for Mobile Radio in the 900 and 1,800 MHz Bands," COST 231 TD (90) 119, Rev. 1, January 1991.

Part IV
Data in Cellular Radio Systems

Chapter 12

Data Services Over Cellular Radio

I. Harris

Vodata, Ltd.

The growth of the cellular network for speech communications has led to the cellular environment being regarded as an extension to the fixed-network environment and, in consequence, has led to the need for it to offer many of the facilities already commonplace in the office or home. Inevitably, the requirement for data communications has been embraced. This chapter assesses the market for data over cellular in general and reviews the methods used to support the service in the U.K. analog cellular networks. Chapter 13 addresses specific issues concerning data in the GSM digital cellular networks, while Chapter 14 discusses the special problems of facsimile transmission.

12.1 THE MARKET FOR DATA IN A CELLULAR RADIO ENVIRONMENT

12.1.1 Introduction

In the PSTN environment, the number of users with a data-only requirement (with the exception of facsimile) is small compared to those using speech. Users requiring high-volume, frequent data transmission invariably use data-specific networks rather than the PSTN because of either performance or attractive tariffing. Widespread use of ISDN is urgently needed on a national and international scale. Its slowness in penetrating the market has led to the continued development of sophisticated modem technology in order to achieve higher data rates and to increase our reliance on facsimile group 3 as a means of communication.

It is estimated that the number of data users in a cellular radio environment is about 1% to 2% of the entire mobile user population. Less conservative estimates put the figure at 5%. What is interesting is that a significant number of data users also require speech.

Therefore data-only requirements tend to be for highly specific and often "unmanned" applications. In such circumstances, however, the provision of a packetized communications link can offer a more cost-effective solution than the connection-oriented cellular approach.

A number of data-specific radio networks have been established with varying degrees of success (e.g., Cognito, Hutchison, Ram, Paknet). A data-specific radio network does not necessarily have to be a mobile network, but once mobility is offered, then full national radio coverage becomes an important requirement. This in turn raises a number of secondary issues, such as the ability of the mobile to receive calls. Such a system requires a location update mechanism in order that the mobile can be quickly located. Network management is a key consideration, together with the need for diverse routing in the event of network malfunction. The cost of providing and maintaining the infrastructures is considerable.

Any data-specific mobile network, therefore, has to compete directly with the cellular operators, who have the advantage in that the network infrastructure, its operation and management, is primarily funded by speech calls that offset the costs involved in providing support for data. Conversely, a data-specific network only has income from data to support an equally complex network. It is hardly surprising, therefore, that data-specific networks face a difficult challenge in the face of cellular operators, except for those applications where there is a substantial cost saving to the user through an alternative approach.

12.1.2 Cellular Data Applications

The application areas, for which data over cellular is an attractive proposition, generally fall into two categories.

- Those involving a speech communication requirement;
- Those required by users who already have a cellular phone for other purposes.

Some examples of cellular data applications in current use are parcel dispatch, retail and wholesale distribution, vehicle wheel-clamping operations, service-engineer support, and sales support.

In a number of these applications, calls were previously being established manually and information being passed by voice. In consequence, call durations were often lengthy, thereby incurring high call charges, and, worse still, errors were being introduced unintentionally during the verbal transfer of information. The use of data for these applications resulted in a dramatic reduction in call duration, and hence operational costs, and eliminated the problem of erroneous transfer of information.

Most data applications have a common requirement to transfer information in an efficient, cost-effective, and reliable manner. To facilitate this, the more cost-conscious users require automation in their terminal equipment or their PC. This usually entails having a software package that not only provides a file transfer protocol, but also provides an interface between the application software and the mechanism for automatic call establishment. However simple this software functionality may be, the degree of standardization leaves much to be desired, and therefore the effort needed for a user to create an operational data environment is a significant obstacle. The problem is not wholly attributed to cellular radio. Until the use of data in the fixed network becomes more widely established and accepted, it is unlikely that data over cellular will make any significant market penetration. The main driving force for cellular data is undoubtedly specific necessity and is therefore an unlikely catalyst to promote a more widespread use of data in the fixed network.

Paknet

Paknet, Ltd., in particular has become well established in a very short space of time, and a brief mention of their radio network seems appropriate here. Paknet was established in 1990 as a joint venture between Vodafone, Ltd., (then Racal Telecom) and Mercury. Paknet's initial aim was to address the EFTPOS market, and it has subsequently made a significant penetration into security applications and remote meter reading. The packet network comprises a VHF radio environment and an X25 backbone fixed network linking the radio base stations to a number of X25 switches and to a network management center [1,2].

User access in the radio environment is achieved by a network termination unit (NTU). The NTU is basically a radio pad and provides the user with two independent V24 asynchronous ports, each of which provides a subset of X28 access procedures. Each port also offers V25bis access procedures for those users who are more familiar with modem access protocols.

The protocol across the air interface is known as Dynamic Slotted Reservation Aloha (DSRA). DSRA is a layer 2 protocol with Golay-encoded FEC. The base station transmits continuously, providing slot synchronization and the opportunity for NTUs to bid for timed access slots. An NTU wishing to transmit will first obtain slot synchronization and then transmit its data at an appropriate time assigned for access by the base station. When the data sent by the NTU reach the base station, the base station sets up an X25 call to the destination requested by the NTU. This may be a fast select type call. The virtual circuit is established between the NTU and the host for the duration of the session. Hosts may also initiate calls to an NTU, since each NTU has a unique address and a registration system ensures that the location of an NTU is known.

Paknet has X121 addressing and its own Data Networking Identification Code (DNIC), which facilitates interworking with other packet-switched networks such as BT's packet switched service (PSS), BT Dialplus).

12.1.3 Impact of GSM

It seems unlikely at present that the GSM cellular environment will generate any significant increase in the use of mobile data in the United Kingdom. It is, however, fair to say that standardization in GSM will make it easier to use data, but the fundamental lack of standardization at the higher layers of data communications remain unchanged, and this is the underlying reason why facsimile group 3 is used so extensively in the fixed network.

The GSM SMS will undoubtedly satisfy the requirements of a significant number of existing and potential mobile messaging-type data applications on cellular radio, as well as those applications currently using dedicated mobile data networks. It is unfortunate that the operators of data-specific networks are now seemingly faced with the challenge of the GSM network, particularly its SMS. It must be appreciated, however, that there are still a number of applications that do not necessarily require speech, and providing these services are operationally satisfactory, properly managed, and attractively tariffed, then they will undoubtedly survive. Chapter 13 specifically describes the data facilities provided by GSM.

12.2 DATA IN THE U.K. (TACS) ANALOG CELLULAR NETWORK

There are two ways in which data transmission may be supported in the analog cellular network.

- Through the use of conventional PSTN modems;
- Through the use of cellular-specific modems.

These two approaches are discussed in Sections 12.2.3 and 12.2.4; but first, it is necessary to understand some fundamental differences between the cellular radio environment and that of the PSTN (a combination of which may be used in a data call).

12.2.1 Transmission Impairment

Data sent by cellular radio is subject to radio transmission impairment that is not present in a conventional PSTN or any fixed-network environment.

Figure 12.1 gives an indication of the nature of this impairment. Thus, obstacles such as buildings, trees, or vehicles in the radio path reduce the signal level at the receiver. A receiver in such a situation is said to be in a radio shadow. The reduction in signal strength may be transitory if the obstacle, the transmitter, or the receiver is moving. This effect is known as fading, whereby the received signal level drops momentarily. The duration of the signal loss may vary from a few milliseconds to several seconds. If the loss exceeds 5 seconds, then the call will be cleared by the network. Figure 12.2 shows the nature of a typical signal envelope in fading and shadowing conditions.

Radio signals deflected from obstacles not necessarily in the radio line of sight arrive at the receiver with varying path differences, causing mutual cancellation of signal level at the receiver. If either the transmitter, the receiver, or the obstacle are moving, the problem can become compounded due to a Doppler frequency shift of the components.

Loss of signals for several hundreds of milliseconds can also occur due to the signaling strategy of cellular radio, such as *power stepping* the mobile as it approaches

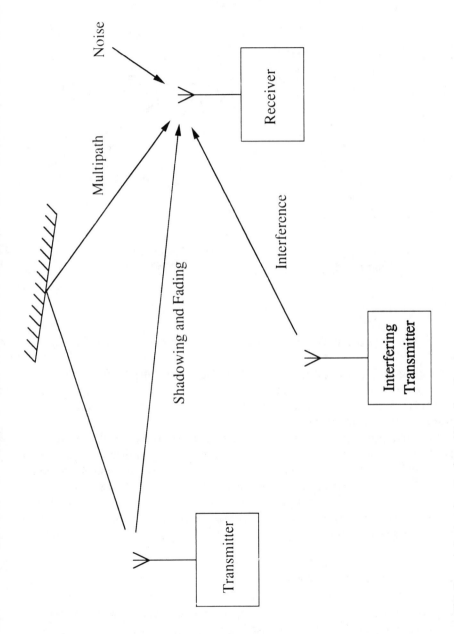

Figure 12.1 Transmission impairments of a mobile radio path.

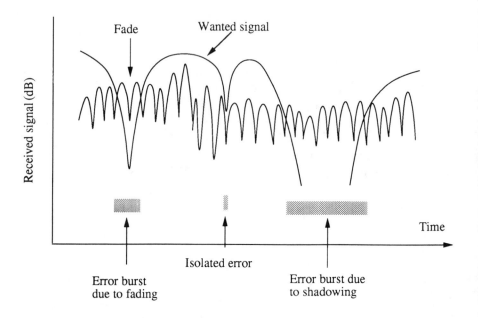

Figure 12.2 Fading and shadowing of the received signal envelope.

or moves away from a cell site, and handover when moving from one cell boundary to another.

If analog data transmission is to be sustained under the above conditions, then clearly some protection is needed to guard against bit errors and burst errors, which result from momentary signal loss, and also against prolonged loss of signal, which may cause loss of the overall carrier signal.

At a data transmission rate of 2,400 bps, the perceived BER at the receiver can be on the order of 2% (i.e., 1 in 50 bits in error). This figure is substantiated by both laboratory simulation and extensive field trials conducted in the United Kingdom. In comparison, one is unlikely to experience PSTN BERs exceeding 1×10^{-5}. Hence, the BER in the cellular radio environment may be 2,000 times worse than that of the associated fixed network. Furthermore, the PSTN does not normally experience a total loss of connection for several hundreds of milliseconds.

12.2.2 Protocols and Error Protection

In any relatively noise-free environment, data communications generally rely upon a layer 2 protocol such as High-Level Data Link Control (HDLC) to safeguard the user's data against errors. Such protocols provide supervision of the data path and can automatically request retransmission of erroneous data. This is known as ARQ and is illustrated in Figure 12.3.

The data to be transmitted are divided up into blocks, each block having a CRC code that allows the receiver to check for errors, plus a header that contains supervisory information, such as a block reference number and the reference number of the last correctly received block. If a block is received with errors or missing data, the receiver requests retransmission and the transmitter retransmits it at the earliest opportunity.

The block reference number is usually a modulus that allows the protocol to "free wheel" up to the modulo value in the event of an erroneous block being received. Under severe bit error conditions, layer 2 protocols fail to function effectively, since requests for retransmission may also in themselves be erroneous. This can cause data transmission to cease because the modulus has been reached before the erroneous block has been successfully retransmitted. This effect is known as *windowing*. If the cessation of data transmission is prolonged, then the layer 2 protocol may initiate a restart procedure. More information on a typical layer 2 protocol may be found in CCITT Rec. X25 [3] or the ISO 4335 Standard for HDLC procedures [4]. In order to minimize the risk of windowing, forward error correction becomes necessary.

12.2.2.1 Forward Error Correction

FEC is a technique whereby erroneous data may be corrected at the receiver, reducing the need for frequent ARQ, or removing the need for ARQ altogether.

The source data is encoded prior to transmission according to an appropriate FEC algorithm, the effect of which is to include additional information. The ratio of the total information content to the user information carried is known as the *FEC code redundancy* and is expressed in the FEC code description. There are basically three common FEC code types: binary block codes, symbol block codes, and convolutional codes. The most common class of symbol block codes are the Reed-Solomon (RS) codes. Convolutional codes tend to be more complex to implement. Binary block codes can be highly effective in correcting for a large number of bit errors, yet are relatively simple to implement; the most common probably being the Bose-Chaudhuri-Hocquenghem (BCH) codes. Invariably, the choice of FEC code will be a tradeoff between implementation complexity, performance, and code redundancy. However, errors tend to occur in bursts, reducing the effectiveness of the FEC code.

12.2.2.2 Interleaving

In order to distribute the effect of burst errors, a technique known as interleaving is used. Instead of transmitting all the bits of one code block together, the bits of a given code block are spread out in time by interleaving bits of other code blocks. The interval between successive bits belonging to the same code block is called the *interleaving depth.*

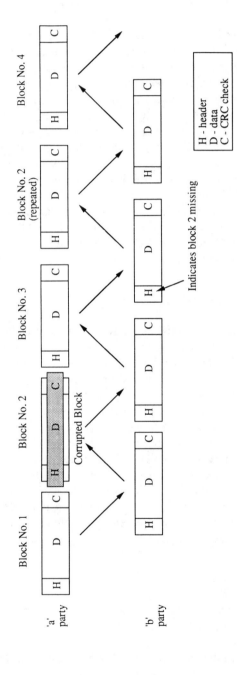

Figure 12.3 Example of ARQ in a full duplex environment.

Table 12.1
Effectiveness of FEC and ARQ

Error Condition	Typical Values	Method of Protection
Isolated errors due to noise	BER <2%	Forward error correction and interleaving
Short bursts due to fading	BER <2% Fades 1–10 ms	Forward error correction and interleaving
Signalling breaks*	100–300 ms	Automatic repeat requests
Radio shadows	Up to 5 seconds[†]	Automatic repeat requests

*For example, handover.
[†]Before call is cleared down.

In Figure 12.4, an error burst occurring in an FEC-encoded noninterleaved bit stream could result in a large number of bits (say 4) in any one character becoming corrupted. The FEC algorithm may only be capable of correcting perhaps up to 2 bits in error in any one block number and would therefore fail. This implies that the layer 2 protocol would invoke an ARQ, if it were available. But when the burst error occurs in an interleaved bit stream, only 2 bits of each block are corrupted (i.e., only 1 bit in each block), and therefore the FEC algorithm is able to correct all the bits in error in each block.

Provided the error burst lengths are less than the interleaving depth, FEC codes can handle burst errors just as well as random errors. Increasing the interleaving depth increases the capability of the code to combat burst errors, but has the disadvantage of increasing the length of the message. In general, the more complex the code and greater the interleaving depth, the greater the transmission delay.

Now that the problems affecting the transmission of data in an analog cellular environment and how they may be overcome have been reviewed, the next section discusses the means by which data transmission may be supported in such an environment.

12.2.3 Conventional PSTN Modems

Conventional modems have been widely used in the PSTN for many years. They have grown ever more sophisticated, offering higher speeds and error correcting protocols. Many variations are in wide use in the PSTN today, from V21 modems offering the user 300-bps full-duplex operation, to V32 offering the user 9,600-bps full-duplex operation with V42 error correction.

A list of commonly used modems and a summary of their characteristics, which are relevant to their performance in a cellular radio environment, is shown in Table 12.2.

The more intelligent higher speed modems usually offer integral layer 2 data protocols. CCITT V42, for example, provides a negotiation procedure for the selection of one of two error correction mechanisms, one of which is the more commonly known MNP and the other is LAPM. The success in market penetration of high-bit-rate modem technology is

Noninterleaved bit stream

Block   A   B   J

Bit no. 1 2 3 4 5 6 7 8 | 1 2 3 4 5 6 7 8 | . . . . | 1 2 3 4 5 6 7 8 |

Interleaved bit stream

A1 B1 . . . . . . J1 | A2 B2 . . . . . . J2 | . . . . | A8 B8 . . . . . . J8 |

↕ Intrerleaving depth

Error burst

Blocks at the receiver

A   B   J

1 2 3 4 5 6 7 8 | 1 2 3 4 5 6 7 8 | . . . . | 1 2 3 4 5 6 7 8 |

Errors

Figure 12.4 Interleaving.

Table 12.2
Modem Characteristics

Modem Type	Bit Rate	Modulation	Equalization	Scrambler
V21	300/300	FSK	No	No
V23	1200/75	FSK	No	No
V26bis	2400/150	QPSK/FSK	Fixed	Optional
V22	1200/1200	QPSK	Fixed	Yes
V22bis	2400/2400	16L-QAM	Adaptive	Yes
V27ter	4800/75	8 phase PSK/FSK	Adaptive	Yes
V32	9600/9600	16/32 phase QAM (Trellis)	Adaptive	Yes

largely attributed to the improvement in the quality of PSTN circuits, as well as to sophistication in modulation coding techniques.

Data users wishing to extend their data applications to the cellular environment often do so by connecting their PSTN modem to their cellular phone through an adapter box. Unfortunately, the performance of PSTN modems then tends to be erratic, primarily because such modems (with or without V42), were never intended to work in a high-BER environment like that of cellular radio. The higher speed modems have complex synchronization methods that do not perform well in high-BER conditions. Also, carrier losses of even a few hundred milliseconds often result in the modems disconnecting from the line.

Half-duplex modems are seldom used today (except in facsimile apparatus). Sustained carrier loss in a cellular environment for half-duplex operation can severely confuse such modems, since they are often unable to distinguish between an intentional loss of carrier, which one half-duplex modem uses to tell the other modem to transmit, and loss of carrier due to poor line conditions. Nevertheless, it would be wrong to say that PSTN modems generally do not work in a cellular environment. In fact, given certain radio conditions at a given point in time, they seem to work quite well. The fundamental problem is that their performance is not consistent across the cellular network and is, at best, erratic. A modem with such a characteristic is clearly a source of frustration to the user and will result in an escalation of call costs due to the need to redial in the event of aborted calls or due to prolonged periods of data retransmission because the equipment cannot cope with the high BER.

Figure 12.5 shows the performance of V23, V22bis, V26, and V29 modems in a static and mobile (20 mph) environment. The cellular network on which these tests were conducted was set up to give a short-term mean carrier-to-noise ratio (CNR) and CIR in excess of 18 dB over 90% of the cell area. Therefore, the 18-dB point on the horizontal axis of the graph is the reference point.

Given the design criteria of a 1 in 50 (2%) BER and an 18-dB threshold, it is apparent that neither the V29 modem at 4,800 baud or the V22bis modem gives adequate performance in these conditions. Furthermore, the V29 and V22bis modems include

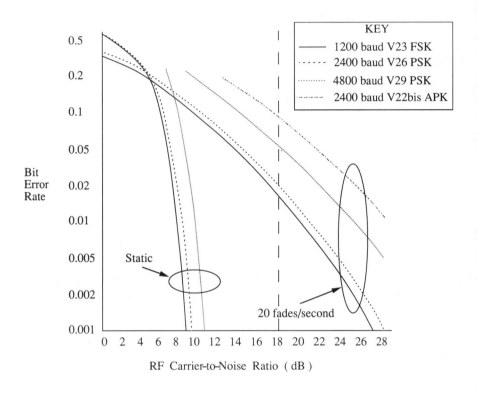

Figure 12.5 Modem performance under cellular conditions.

scramblers to improve the performance of the demodulator. Polynomial descramblers have a tendency to propagate errors (error multiplication) when receiving data from a high-BER environment.

V29 and V22bis modems require a wider bandwidth and are thus more affected by amplitude and phase distortion incurred in the audio processing stages of a cellular transceiver.

Fading was also found to cause the adaptive equalizers to malfunction. The performance would worsen by some 2 dB when equalization was included, primarily because of attempts by the equalizer to adapt to an unstable line condition.

In so far as V21 and V23 modems are concerned, the user data rate they are capable of providing would not be acceptable for the majority of applications.

Given the considerations described above, the development of a cellular-specific modem and protocol within Vodafone concentrated on V26 (see Section 12.2.4).

12.2.4 Cellular-Specific Modems

Modems specifically designed for cellular radio will invariably have addressed the problem of data transmission in such an environment. One such product is the cellular data link control (CDLC) modem, which incorporates a CDLC protocol. This modem can interface to different cellular phones. Since phone manufacturers did not agree on a common interface standard, either physical or operational, the CDLC modem is supplied with an appropriate interface cable and is fitted with phone-specific software. It can be supplied with a remote terminal interface (RTI), which provides remote voice, data, and auto answer selection and remote terminal access, allowing the modem to be installed in any convenient place in the vehicle. It has an asynchronous terminal interface conforming to CCITT V24/V28 standards. The physical interface is a 25-way D-type connector.

Experience has shown that users of data communications equipment generally have very little knowledge of data communications processes (e.g., data rates, network tones, parity), and consequently it is desirable to minimize the need for users to operate at this level of detail. Vodafone, sympathetic to this fact, developed the modem with auto call/ auto answer facilities according to a subset of CCITT V25bis. This enables wholly automated application packages, which run on PCs, to be written by third parties who are knowledgeable about data communications. More important still, automating the line connection and disconnection procedure ensures that calls are not held for longer than is necessary. (It is possible with the CDLC modem to optionally manually establish and terminate a call.)

Most data calls are initiated by the mobile. A mobile is rarely available to answer calls, and even if it were, the likelihood of having the correct data equipment available is remote. Were it not for the use of voice message storage facilities, even mobile-terminated speech calls would be virtually unusable. For data, not only are data "store and forward" systems not commercially viable or operationally desirable, but a call arriving at a mobile is not known to be a data call until it has actually been answered. Unless data equipment is permanently available to answer the call at the mobile station, the operational constraints on a user are obvious. The majority of mobile-terminated data calls are for special data-only applications.

12.2.4.1 CDLC Protocol

CDLC [5] is a full-duplex layer 2 protocol in the public domain and is based on HDLC standards. It retains all the procedural elements of HDLC, including "selective reject" for more efficient error recovery. The means of frame synchronization has been modified in order to cope with high BERs, which would otherwise have a catastrophic effect on HDLC flags. CDLC also includes BCH and RS FEC codes and interleaving.

The development of the CDLC protocol and subsequently the CDLC modem itself was the result of almost a year's extensive field and laboratory trials during 1985. These trials were divided into a number of identifiable tasks:

- Assessing the severity of problems such as fading and handover;
- Assessing the performance of PSTN modems in the cellular environment; and
- Assessing the performance of FEC codes in the cellular environment.

12.2.4.2 CDLC Modulation

The assessment of the performance of PSTN modems was primarily to establish which modem standard could be adopted for error-free data transmission over cellular radio. The results indicated that V26 provided an acceptable user data rate of 2,400 bps and adequate resilience. It is not permissible to use a V26 modem in its full-duplex mode on the PSTN because of its spectral occupancy requirement, but the provision of network data services (see Section 12.2.5) obviates the need to do so.

However, in those environments outside the United Kingdom where network data services may not be possible to implement, the performance of the backward channel of V26bis was assessed at 150 bps with the view to providing the capability of interworking directly with the PSTN. The use of the backward channel enables full-duplex operation on the PSTN with certain limitations, which are discussed later. It should be stressed that, in the U.K. environment, the backward channel in the CDLC modem is rarely used because of the network data services provided by Vodafone.

12.2.4.3 CDLC FEC Codes

CDLC uses a 16,8 BCH code and an RS 72,68 code whose performance is shown in Figure 12.6. The code chosen for the backward channel was a 23,12 Golay code.

The RS code chosen, while weaker than the BCH code, is automatically enabled when the perceived error rates are low. When the error rate increases, the BCH code is automatically selected and remains operational until the error rate decreases. The algorithm for switching between the two is quite complex in order to avoid instability, and a description of it is outside the scope of this publication. One other feature is that if an ARQ is necessary when using an RS code, then the retransmitted block is always sent BCH-encoded. The longest BCH-encoded CDLC frame gives an efficiency of over 55% and, therefore, using a 2,400-baud modem, a user data rate of 1,200 bps can be achieved, even if 10% of blocks have to be retransmitted in event of FEC failure. Similarly, the RS code will support a user data rate of 2,400 bps with allowance for repeat transmissions.

Throughput can only meaningfully be expressed in terms of the data rate as perceived by the user. Clearly, in the cellular radio environment, this is a function of the error rate. Using a cell design threshold signal strength for acceptable voice communications (−110 dBm), CDLC will support a user full-duplex data rate of 2,430 bps over typically 80% of the cell area for RS FEC and 1,340 bps over typically 90% of the cell area for BCH FEC. It should be noted that these data rates allow for the fact that start/stop bits of asynchronous DTEs are stripped out and are not transmitted by CDLC.

The performance of FEC codes applied to a block of data relative to vehicle speed and cell signal strength is shown in Table 12.3. Any percentage of block errors greater than zero would result in ARQ.

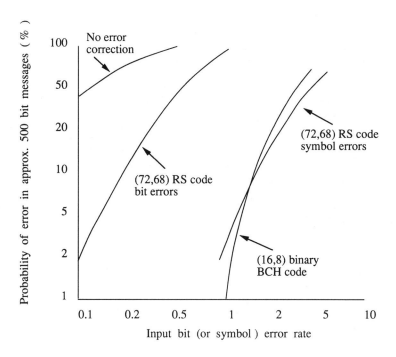

Figure 12.6 Performance of FEC codes.

12.2.4.4 CDLC Synchronization.

The 8 bit flags used in HDLC for frame delimiting and synchronization are not error-protected; therefore, a 48-bit error-protected leading synchronization code is used in CDLC. In order to avoid the need to repeat this 48-bit synchronization code as a trailing flag, CDLC has in fact eight different leading synchronization codes, each code defining one of eight frame lengths, thereby obviating the need for a trailing flag. These different frame lengths also enable data to be transmitted with minimal delay in those situations where only a few user data characters are to be transmitted together. See Table 12.4.

The synchronization codes were chosen to give a low probability that random data could be mistaken for one of them, but a high probability of successful synchronization in the presence of errors. Up to eight bit errors can be tolerated in any one synchronization code. The probability of a random data stream being mistaken for one of the eight codes is 9.1×10^{-6}. The probability of successful synchronization with a random BER of 2% is 99.96%. The synchronization code is not FEC-encoded or interleaved with the user data.

Table 12.3

Field Trial Results

Physical Conditions		2400 baud channel, 512 data bits per block		150 baud channel, 36 data bits per block	
		Percentage Block Errors (after FEC)			
Speed (kph)	*Signal Strength (dBm)*	*No FEC*	*Interleaved 16,8 Code (BCH)*	*No FEC*	*Interleaved 23,12 Code (Golay)*
0 to 20	−120 to −110	23.7	0.6	8.8	0.0
0 to 20	−110 to −100	17.5	0.6	5.3	0.0
0 to 20	−100 to −90	11.1	1.0	5.4	0.0
0 to 20	−90 to −80	8.3	0.0	4.5	0.0
20 to 40	−120 to −110	49.8	0.0	24.2	0.0
20 to 40	−110 to −100	42.6	0.0	23.1	0.0
20 to 40	−100 to −90	27.8	0.2	11.7	0.0
20 to 40	−90 to −80	18.1	0.0	4.4	0.0
40 to 75	−120 to −110	58.5	1.7	—	—
40 to 75	−110 to −100	36.1	0.0	—	—
40 to 75	−100 to −90	29.2	0.6	4.2	0.0
40 to 75	−90 to −80	17.8	0.0	2.0	0.0

Table 12.4

CDLC Synchronization Codes

Sync Type	*Block Length (ms)*	*Interleaving Depth (ms)*	*Information Characters in Block (N)*
0	47	1.7	0
1	60	2.5	2
2	73	3.3	4
3	100	5	8
4	153	8.3	16
5	260	15	32
6	473	28.3	64
7*			

*Uses noninterleaved Reed-Solomon (72,68) code.

12.2.4.5 CDLC Frame Structure

A comparison between the more familiar HDLC frame structure and that of CDLC is shown in Figure 12.7. The 48-bit CDLC synchronization code is not included in the CRC check sum.

291

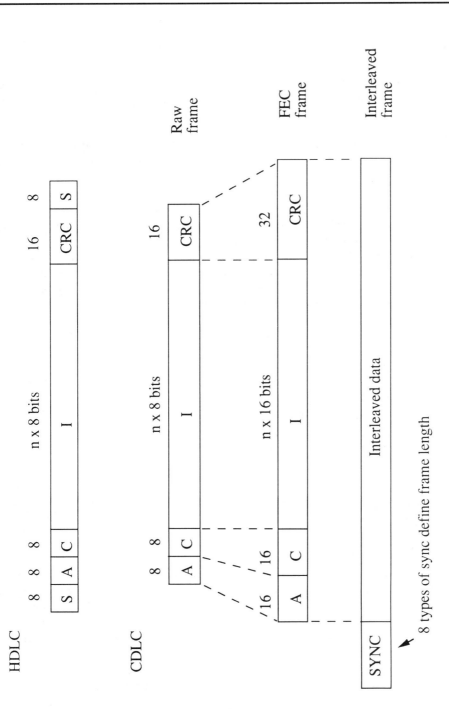

Figure 12.7 Comparison of HDLC and CDLC frames.

12.2.5 Data Services in the Analog TACS U.K. Network

Data services can best be defined as the ability of a network to support the interworking between the cellular environment and the fixed-network environment (e.g., PSTN, PSPDN).

Vodafone's approach emphasized the need to decouple the standards of the fixed networks from those of the cellular radio environment, and to this end it implemented a data gateway to provide the interworking between a mobile and the fixed network. The data gateway is the Vodafone Mobile Access Conversion Service (VMACS). The philosophy of such gateways has been adopted by GSM, where it is known as an *interworking function* (IWF). Instead of CDLC, GSM has defined a Radio Link Protocol (RLP) to protect data across the radio path. Both CDLC and RLP protocols have common roots in HDLC and incorporate FEC and interleaving.

The approach to interworking adopted by Cellnet in the U.K. TACS network was to let the user choose the type of modem for the mobile station and, although some data gateways were made available, direct interworking between the mobile and, for example, the PSTN apparatus was more widely encouraged.

12.2.5.1 VMACS

The Vodafone VMACS data gateway provides interworking between CDLC with the following fixed networks:

- PSTN—V21, V22, V22bis, V23, V42 error correction;
- PSPDN—BT, Mercury, IBM, ISTEL;
- Private networks.

The mobile user can access any of these services using the same CDLC modem. No charge is made for the use of VMACS. All services offer full-duplex bidirectional simultaneous data transfer. See Figure 12.8.

12.2.5.2 PSTN Interworking

The mobile user often has no knowledge of the type of modem used by the PSTN number he or she wishes to access. Therefore, in the VMACS service, it is arranged that the mobile user merely needs to prefix the PSTN subscriber trunk dialing (STD) code and *b* number with 972. The PSTN number is routed directly to the PSTN (without the need for two-stage dialing). When the *b* party answers, the PSTN modem in VMACS adapts itself to the modulation and error correction standards dictated by the answering *b* party modem. At the same time, CDLC establishes itself across the radio path.

The data rates set by the mobile data termination equipment (DTE) are independent of those used in the PSTN. Out-of-band flow control using the V24 control signal clear-

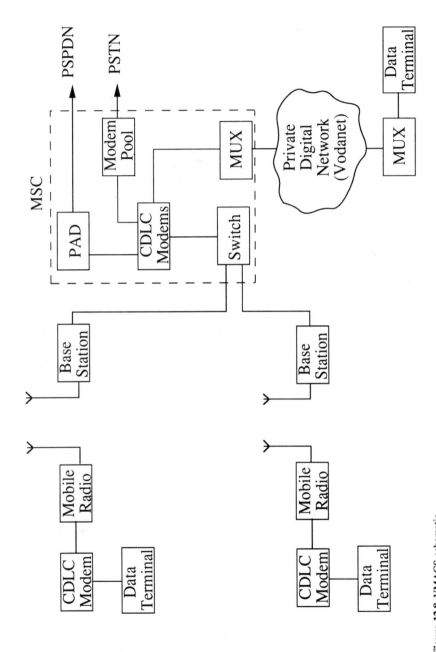

Figure 12.8 VMACS schematic.

to-send (CTS) is provided by the data gateway so that the service is transparent to any inband flow control characters such as X-ON and X-OFF (DC1 and DC3) in the CCITT IA5 character set, which may be used by the DTE.

Vodafone has recently introduced higher speed signaling within their network for the 972 service, which has resulted in an improvement in the time taken to establish connection. See Figure 12.9.

12.2.5.3 PSPDN Access

Mobile access to packet switched networks is provided by dialing 970XXX, which connects the mobile to a packet assembler/disassembler (PAD) on the appropriate network, defined by the digits XXX. The PAD belongs to the PSPDN, and invariably, the mobile user has to be supplied with a password to access any of the hosts supported by the PSPDN. Conventional PSPDN logon and host access procedures apply. See Figure 12.10.

12.2.5.4 Private Networks

These are ideally targeted at the large corporate users, where the cost of providing such a network is justified by reason of lower call charges. Vodafone offers Vodanet users access to their private data network with the 973 service. This service uses digital multiplexers and 64-Kbps channels within one 2-Mbps Vodanet private circuit connection. Up to 24 simultaneous data calls can be in progress on one 64-Kbps channel. Access is achieved by merely prefixing calls to private network access codes with 973. The service also allows users in Vodanet to make calls to mobiles. See Figure 12.11.

12.2.5.5 Reverse PSTN VMACS

For users in the PSTN wishing to access a mobile, there is no convenient way in which a mobile access code (e.g., 0836) can include discrimination between a data call and a voice call. VMACS offers the 0836 331441 service. This requires two-stage dialing; that is, when access to the 331441 service is established, the PSTN calling party is then required to enter the mobile number. The service provides interworking with the PSTN for V21, V22, V22bis, V23 modems with or without PSTN error correction in a way similar to the 972 service.

12.2.5.6 Test Services

The VMACS services described above are provided with a test service number to allow mobile users to check the operation of their mobile equipment. This test service has a number of facilities, defined by appending appropriate digits to the end of 970123 when

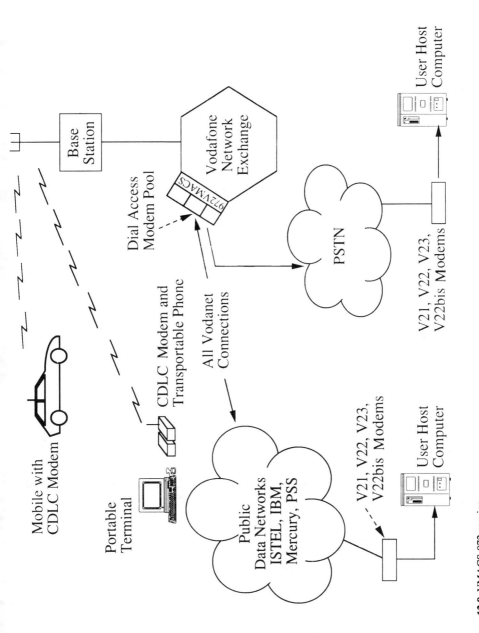

Figure 12.9 VMACS 972 service.

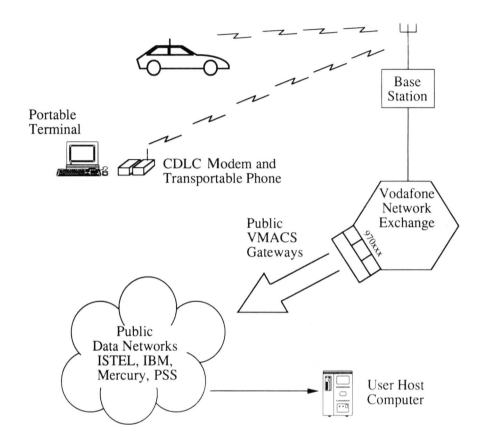

Figure 12.10 VMACS 970XXX services.

dialing. For example, 97012311 will provide a menu of facilities offered, which includes a one-minute transmission of scrolling characters or dialback.

12.2.5.7 VMACS Management

The management of VMACS is an integral part of the operational strategy, ensuring that faulty equipment or circuits can be automatically removed from service in the event of malfunction and an appropriate alarm raised. Frequent traffic monitoring is part of this strategy, allowing observation of potential traffic congestion and statistical analysis in order to establish the requirement for increasing service capacity. It also allows an insight into the user's operational demands (e.g., call holding times). Diversity is also provided

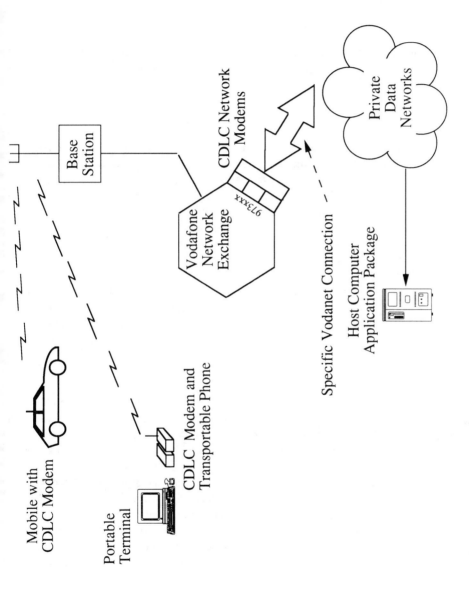

Figure 12.11 VMACS 973 service.

by ensuring that common hardware failure causes minimal disruption to service availability, and, wherever possible, diverse routing is applied.

12.2.5.8 Mobile-to-Mobile Calls

CDLC modems are capable of mobile-to-mobile operation using the symmetric full-duplex mode of operation (i.e., 2,400 bps in both directions simultaneously). These calls do not go via VMACS.

12.2.5.9 Direct Interworking With the PSTN

The CDLC modem is capable of directly interworking with the PSTN without going via VMACS. As mentioned earlier, this mode of operation is less applicable in the United Kingdom. CDLC can be selected to operate in an asymmetric full-duplex mode, sometimes referred to in CDLC as the *2-wire* mode. The forward channel operates at 2,400 bps and the backward channel at 150 bps. This enables the full-duplex CDLC protocol to perform as if a symmetric full-duplex operation (2,400 bps both directions) were being used.

The main limitation is that mass data transmission simultaneously in both directions is not possible. The direction in which the 2,400-bps data operates is automatically selected according to the direction in which most data are flowing. This may subsequently be reversed by a line reversal code, which is heavily coded to protect against errors. Simple interactive Videotext-type applications are possible, with no degradation in performance, since a few characters will not invoke a line turn-around even if there is no data flowing in the opposite direction.

12.3 SUMMARY AND CONCLUSIONS

It can be said that in the United Kingdom there has been an element of confusion concerning data over cellular radio. Apart from the choice between CDLC and conventional PSTN standards, the benefit of data gateways also has to be considered. The two operators in the United Kingdom have taken different approaches with the provision of data services, and this has given users a freedom of choice.

Vodafone's approach is that the mobile user need not be aware of the various standards used in the fixed network and that it is beneficial to have just one mobile CDLC modem to access any service. Cellnet is supportive of data gateway concepts and of the protection provided by CDLC, but their advocacy of conventional PSTN modems for the cellular environment makes them give the need for one standard for the mobile environment and for cellular-specific data gateways a somewhat lower profile.

REFERENCES

[1] Davie, M. C., I. Harris, and J. B. Smith, "A Cellular Packet Radio Network," *IEE Conference on Radio Receivers and Associated Systems,* Churchill College, Cambridge, England, July 1990.

[2] Davie, M. C., and J. B. Smith, "A Cellular Packet Radio Data Network," *Electronics and Communications Engineering Journal,* Vol. 3, No. 3, June 1991, pp. 137–143.

[3] CCITT Recommendations X25, International Telegraph and Telephone Consultative Committee, Geneva.

[4] ISO 4335, Elements of Procedures, International Organization for Standards, Geneva.

[5] Frazer, E. L., I. Harris, P. Munday, "CDLC: A Data Transmission Standard for Cellular Radio," *Journal of the IERE,* Vol. 57, No. 3, May/June 1987, pp. 129–133.

FURTHER READING

Bleazard, G. B., *Handbook of Data Communications,* Manchester: NCC Publications, 1982.

Jennings, F., *Practical Data Communications, Modems, Networks and Protocols,* Blackwell Scientific Publications, 1986.

Brewster, R., ed., *Data Communications and Networks 2,* London: Peter Peregrinus, 1989.

Holbeche, R. J., ed., *Land Mobile Radio Systems,* London: Peter Peregrinus, 1985.

Chapter 13
Data in the GSM Cellular Network

I. Harris
Vodata, Ltd.

The digital nature of GSM and its architectural basis in ISDN means that there are no special requirements for conventional modems when using the service for data transmission. It is, however, necessary to understand how data are treated by the network in order to select the most appropriate of the several methods available for any specific application. This chapter describes the different ways that data can be transmitted by a GSM user and outlines the data conversion processes that GSM introduces.

13.1 INTRODUCTION

The architecture of the GSM infrastructure is based on ISDN and there is therefore no requirement for a special modem technology except to support interworking with the PSTN. The interface at the mobile station is digital, and digital bearer services are used to transport data across the PLMN to data gateways known as the *interworking functions*, which interface with the various fixed networks. This chapter describes the basic data structures used by GSM, explains how the data rates relate to the standard ISDN rates, and discusses the various bearer and teleservices that are offered.

13.2 BASIC DATA STRUCTURE

Up to eight simultaneous data traffic transmissions may be multiplexed onto a single radio frequency carrier using TDMA. This condition is known as the full-rate traffic channel. To economize on bandwidth for lower data rates, up to sixteen simultaneous transmissions per carrier are possible. This condition is known as the half-rate traffic channel. Data channels and speech channels can be mixed on a single carrier, since their time slot structures are compatible

In order to protect data transmissions against the high BERs of the radio environment, FEC and interleaving is an integral function of the bearer service. Table 13.1 shows the convolutional codes applied to support various user data rates and the delay that can be expected due to the interleaving block length.

The digital nature of GSM necessitates a number of data rate conversions. Figure 13.1 illustrates how data is supported within GSM. A data adapter in the mobile converts the user DTE data rates into 3.6, 6, or 12 Kbps.

FEC and interleaving is then applied to give a data rate of 22.8 Kbps (full rate) or 11.4 Kbps (half rate) for transmission in 8 or 16 TDMA frames, respectively. The base station then converts the 22.8/11.4-Kbps data rate through the FEC function to standard ISDN rates of 16 or 64 Kbps, depending on the type of connection between the base station and the IWF. Half-rate channels may eventually be offered on a differential tariffing basis to achieve better spectral efficiency, but at the expense of performance.

13.2.1 Data Adapter

The data adapter in the mobile has the task of converting the data rates of the DTE to those required by the FEC encoder/decoder. A similar function is necessary at the IWF. This is achieved by using CCITT V110 rate adapters similar to those used in the ISDN. The RA0, RA1, and RA2 functions are standard. The RA1′ function has been designed specifically for GSM, since it was necessary to satisfy the interface rates of 3.6, 6, and 12 Kbps of the FEC encoder/decoder. Figure 13.2 shows how the RA0 and RA1′ rate adapters may be used in the data adapter shown in Figure 13.1. The RA0 function converts asynchronous data rates to an equivalent synchronous data rate and vice versa. The RA1 function converts the synchronous data rates of the RA0 function to the standard ISDN intermediate rates of 8 or 16 Kbps and vice versa. The RA2 function similarly converts

Table 13.1
FEC and Interleaving

Data Rate	Channel	Convolutional Code Redundancy	Interleaving Block Length (ms)
9.6 Kbps	Full rate	61/114 = 0.53	95
4.8 Kbps	Full rate	1/3	95
4.8 Kbps	Half rate	61/114	190
<2.4 Kbps	Full rate	1/6	40
<2.4 Kbps	Half rate	1/3	190
Speech	Full rate	Variable 1/2 or 1	40
Signaling		1/2	20/40

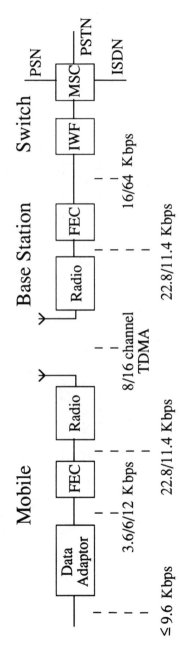

Figure 13.1 Data in the GSM environment.

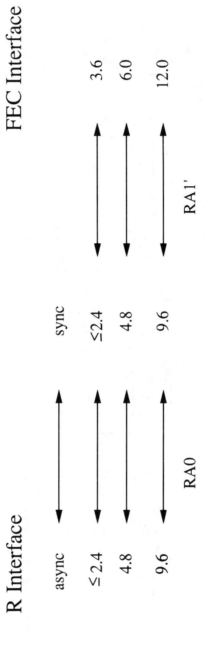

Figure 13.2 Rate adaptation using RA0 and RA1'.

the ISDN intermediate rates of the RA1 function to 64 Kbps. Figure 13.3 shows how all the rate adapters may be used in the data adapter shown in Figure 13.1. The "S" interface may also be regarded as the link between the base station and the switch in Figure 13.1.

13.3 BEARER SERVICES

13.3.1 Transparent and Nontransparent Services

The transparent service provides a basic forward error corrected and interleaved transport mechanism and is capable of supporting both synchronous and asynchronous data terminals. The service has a fixed delay of a few hundred milliseconds end to end, but no guarantee of data integrity. The projected residual error rate (after FEC and interleaving) suggests that the application would normally need to provide a higher layer protection against errors. Typical projected residual BERs for this service (after FEC and interleaving has been applied) are shown in Table 13.2. It should be noted that 9.6 Kbps is not possible on a half-rate channel (HRC) and that the residual BER for 4.8 Kbps on a HRC is the same as that for 9.6 Kbps on a full-rate channel (FRC).

The alternative nontransparent service can support both synchronous and asynchronous terminals and provides an ARQ protection mechanism, RLP, which overlays the transparent bearer and has a negligible residual BER.

RLP provides data integrity at the expense of a connection with variable delay (which could be several seconds in a severe radio fade). It is possible to guarantee user throughput under defined error rates, since the RLP is designed to ensure that the application is unaware of the variable delay.

13.3.2 V110 Rate Adaptation in the Transparent Service

V110 frames normally comprise 80 bits. In order to satisfy the data rate requirements of the FEC function, the V110 frames are modified to 60 bits by removing the 17 synchronization bits and 3 E bits (E1, E2, E3).

At an RA1 rate of 9.6 Kbps, the 60-bit frame (which contains 48 data bits) are sampled every 5 ms, giving a bit rate to the FEC coder of 12 Kbps. At an RA1 rate of 4.8 Kbps, the 60-bit frames (containing 48 data bits) are sampled every 10 ms, giving a bit rate to the FEC coder of 6 Kbps. For data rates of 2.4 Kbps, 1.2 Kbps, and 600 bps, the 60-bit frames are further modified to 36 by reducing the number of data bits to 24. The 36-bit frames are sampled every 10 ms, giving a bit rate to the FEC coder of 3.6 Kbps.

In all the above cases, the 17 synchronization bits are not necessary because synchronization is achieved by other means. In ISDN, the E bits normally convey the data rate information, but in the GSM network the data rate is conveyed by the bearer capability information element (BCIE), which is sent on the signaling channel, when the call is

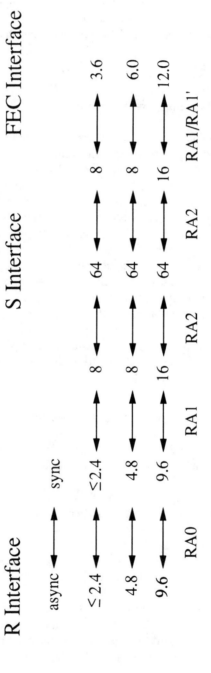

Figure 13.3 Rate adaptation using all rate adapters.

Table 13.2
Typical Bit Error Rates

User Data Rate	Residual BER (Full-Rate Channel)
9.6 Kbps	10^{-2} to 10^{-3}
4.8 Kbps	10^{-3} to 10^{-4}
2.4 Kbps	$>10^{-5}$

established. Table 13.3 shows the difference between an 80-bit V110 frame and a modified V110 frame containing 48 data bits.

13.3.3 V110 Rate Adaptation in the Nontransparent Service

For the nontransparent service, only FEC function rates of 6 and 12 Kbps are used. As mentioned earlier, RLP provides the layer 2 protection for data sent in the nontransparent service. An RLP frame comprises 240 bits (4×60-bit modified V110 frames) and is sent to the FEC function every 20 ms, giving an input rate of 12 Kbps. This is capable of supporting a user data rate of 9.6 Kbps, the difference being absorbed by the RLP overheads and transport of V24 status signals for V series DTEs. In the case of X series DTEs, the correspondence between the X30 frame and a V110 frame is shown in Table 13.4.

13.3.4 X25 Access

Provision is made for X25 DTEs to be attached to the mobile. The transparent or nontransparent service may be used. In the latter case, using RLP, the X25 LAPB is terminated in the mobile station and only the packet layer is conveyed across the air interface in the RLP frames.

13.4 MOBILE STATION

The mobile phone may be one of three types as shown in Figure 13.4. The terminal adapter has to provide RA0, RA1, and RA2 rate adaptation functions in the case of an MT1, whereas in the case of an MT2, the RA2 function may be omitted (see Figures 13.2 and 13.3). An MT0 is a fully integrated digital station which contains the functionality of a data terminal. An MT1 provides an S (ISDN) interface to allow the attachment of ISDN computer terminal equipment (TE1). An MT2 provides an "R" interface to allow the attachment of V or X series (asynchronous or synchronous) terminal equipment (TE2).

Table 13.3
Comparison of V110 Frame to Modified V110 Frame

Octet Number	Bit Number							
	1	2	3	4	5	6	7	8
A. V110 80-bit frame								
0	0	0	0	0	0	0	0	0
1	1	D1	D2	D3	D4	D5	D6	S1
2	1	D7	D8	D9	D10	D11	D12	X
3	1	D13	D14	D15	D16	D17	D18	S3
4	1	D19	D20	D21	D22	D23	D24	S4
5	1	E1	E2	E3	E4	E5	E6	E7
6	1	D25	D26	D27	D28	D29	D30	S6
7	1	D31	D32	D33	D34	D35	D36	X
8	1	D37	D38	D39	D40	D41	D42	S8
9	1	D43	D44	D45	D46	D47	D48	S9
B. Modified CCITT V110 60-bit frame								
0	0	0	0	0	0	0	0	0
1	1	D1	D2	D3	D4	D5	D6	S1
2	1	D7	D8	D9	D10	D11	D12	X
3	1	D13	D14	D15	D16	D17	D18	S3
4	1	D19	D20	D21	D22	D23	D24	S4
5	1	E4	E5	E6	E7	D25	D26	D27
6	1	D28	D29	D30	S6	D31	D32	D33
7	1	D34	D35	D36	X	D37	D38	D39
8	1	D40	D41	D42	S8	D43	D44	D45
9	1	D46	D47	D48	S9	D1	D2	D3

13.5 RADIO LINK PROTOCOL

The RLP is derived from the balanced class of HDLC procedures as defined in ISO 4335 and has been tailored to meet the specific needs of GSM. It has been designed to provide a user date rate of 9.6 Kbps without significant loss of throughput or delay. It is a full-duplex level 2 protocol allowing simultaneous bidirectional transfer of user information. Typically, it is specified that the maximum throughput will be maintained over 90% of the cell area for 90% of the transmission time, allowing for an RLP frame repeat rate (ARQ) of 10%. The elements of procedure in ISO 4335 have largely been preserved, although the frame structure has been modified to make best use of the available channel bandwidth.

A modulus of 64 has been chosen in order to ensure that the round-trip delay, which is on the order of several hundred milliseconds, does not result in windowing and cause data transmission to be temporarily suspended. The ISO 4335 modulus of 128 would

Table 13.4
Correspondence between V110 and X30 Frames

Octet Number	Bit Number							
	1	2	3	4	5	6	7	8
X30 Two-frame Multiframe								
Odd frame 0	0	0	0	0	0	0	0	0
1	1	P1	P2	P3	P4	P5	P6	SQ
2	1	P7	P8	Q1	Q2	Q3	Q4	X
3	1	Q5	Q6	Q7	Q8	R1	R2	SR
4	1	R3	R4	R5	R6	R7	R8	SP
Even frame 5	1	E1	E2	E3	E4	E5	E6	E7
6	1	P1	P2	P3	P4	P5	P6	SQ
7	1	P7	P8	Q1	Q2	Q3	Q4	X
8	1	Q5	Q6	Q7	Q8	R1	R2	SR
9	1	R3	R4	R5	R6	R7	R8	SP
V110 80-bit frame								
Odd frame 0	0	0	0	0	0	0	0	0
1	1	D1	D2	D3	D4	D5	D6	S1
2	1	D7	D8	D9	D10	D11	D12	X
3	1	D13	D14	D15	D16	D17	D18	S3
4	1	D19	D20	D21	D22	D23	D24	S4
Even frame 5	1	E1	E2	E3	E4	E5	E6	E7
6	1	D25	D26	D27	D28	D29	D30	S6
7	1	D31	D32	D33	D34	D35	D36	X
8	1	D37	D38	D39	D40	D41	D42	S8
9	1	D43	D44	D45	D46	D47	D48	S9

have been wasteful of capacity available within the RLP 240-bit frame and would have prevented 9.6 Kbps from being supported. The 16-bit FCS of ISO 4335 gives inadequate protection against the hostile cellular environment, and so a 24-bit FCS was chosen, rather than the 32-bit ISO 4335 alternative, which would again have been wasteful of capacity available within the RLP frame and would have precluded 9.6 Kbps.

RLP has no need for addresses. It also has an optional provision for piggybacking user information with layer 2 control/status information. The simultaneous transmission of these two information types allows more efficient use of the RLP frame, reducing the need for purely supervisory frames during an information transfer and increasing the probability of sustaining data throughput under adverse radio conditions.

An RLP frame is sent in strict alignment with the radio transmission every 20 ms for a FRC and every 40 ms for a HRC. The maximum user data rate supported on a FRC is 9.6 Kbps and 4.8 Kbps on a HRC.

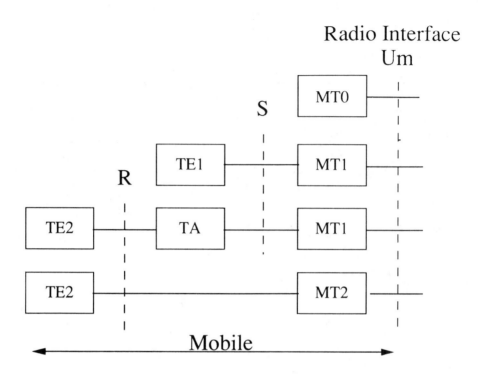

Figure 13.4 Mobile station configurations.

13.5.1 Frame Structure

The RLP frame shown in Figure 13.5 comprises 240 bits and aligns with 4 complete 60-bit V110 frames. The header contains supervisory information, and a full description of RLP will be found in GSM Rec. 04.22

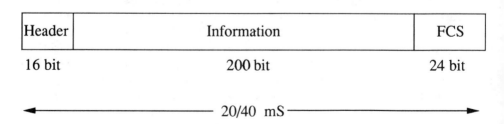

Figure 13.5 RLP frame structure.

13.6 TELESERVICES IN THE GSM DIGITAL NETWORK

GSM supports a number of teleservices:

- Facsimile,
- Videotex,
- Teletex,
- Short-message service, and
- Cell broadcast.

The facsimile teleservice, which needs to be put into context with the provision of facsimile in the U.K. TACS network, is described separately in Chapter 14.

13.6.1 Videotex

Videotex terminals can be supported in a number of ways at the mobile station, depending on whether the Videotex terminal has an analog (2-wire PSTN) or R (V24) interface. Clearly, the preference would be for a Videotex terminal with a V24 interface. Videotex is an interactive service and the effect of delays due to RLP ARQ can have some unfortunate operational implications. For example, if sending a request for a new page is delayed or a partial page is received, the user might enter another command, which will of course be queued, resulting in unwanted interaction a few seconds later.

Videotex data is asymmetric by nature; that is, 1,200 bps from host to terminal and 75 bps from terminal to host (V23 modem standards). The transparent or nontransparent bearer may be used to convey data. However, high BERs in the radio environment can have a dramatic effect on the operation of Videotex, since noise not only severely affects the quality of data displayed on the Videotex terminal, but may also cause spurious prompts for new data. Thus, the use of the nontransparent service seems highly desirable. Unfortunately, it is not possible to flow-control Videotex terminals, and, therefore, buffering for at least one page of Videotex data must exist in the mobile TA and IWF to cater for RLP ARQs.

Provision is made in the IWF to interwork with Videotex hosts connected to the PSTN, ISDN, PSPDN or with direct attachment to the IWF. In view of the above operational difficulties, it seems preferable to use the conventional PSTN access using the nontransparent bearer and a conventional DTE at the mobile rather than a Videotex terminal.

13.6.2 Teletex

Teletex is little used in the United Kingdom, but it does provide the capability of interworking with Telex via a national fixed network conversion facility. Teletex uses a Teletex document store in the fixed network, which provides temporary storage for documents sent to mobiles that may be unavailable. Provision is made in the IWF to interwork with the ISDN, PSTN, CSPDN, and PSPDN.

13.6.3 Short Message Service

The short message service (SMS) allows alphanumeric text messages of up to 160 characters to be sent to and from a GSM phone via a short-message *service center* (SC), irrespective of whether a speech call is in progress. The SC may provide access to fixed networks—such as PSTN, PSPDN, or ISDN—or to equipment dedicated to a particular application. The point of origin and destination of a short message is a short-message entity (SME). An SME may be a GSM mobile phone or a device outside the PLMN, for example, a terminal connected to an X25 network. The prime function of the SC is to store short messages sent from an SME and then attempt to deliver them. The SMS is essentially a store-and-forward service and does not maintain a circuit connection between the origination and destination SMEs.

Short messages received on a GSM mobile phone normally will be stored in the subscriber identity module (SIM card), which is a feature of all GSM phones. It allows a user to remove from the phone their personal identity (their phone number) and any short messages they may have received. A typical SIM currently has storage capacity for five short messages. Phone manufacturers also may choose to store short messages in nonvolatile memory in the phone itself. If another user inserts their own SIM, however, access to short messages that are not their own is prevented. Short messages stored in the phone or on the SIM may be read on the phone's display, which typically is four lines of twelve characters. The short message is accompanied by other information that may be of interest—the originating data address and the time and date received, for instance.

Sending short messages from a mobile phone is complicated by the fact that normally only a ten-digit keypad is provided. The short message must also be accompanied by other information that the user may have to provide, such as the destination data address. (Other attributes, such as the SC data address, will probably be inserted automatically by the phone itself.) The user also may request a status report so that the SC will inform the mobile upon the successful delivery of a message. The limitations of the phone's display area and keypad may be overcome by attaching a computer terminal to the R interface, which allows a user to display and send short messages.

SMS offers a number of other features for sending and receiving short messages.

- Once it has sent a short message to the SC, a mobile may obtain an updated status report or delete or replace a short message.
- When a mobile's memory capacity is full, messages may be kept at the SC; as soon as it has memory available, the mobile may notify the SC for the delivery of any short messages awaiting delivery.
- The SC can replace an already-sent message in a mobile phone, provided that the message is still present in the phone or on the SIM. This allows messages that convey

basically the same information to be updated without using additional memory in the mobile and particularly in the SIM.

- The SC can indicate that a mobile should display a short message immediately rather than store it in the SIM and wait for the user to request display.
- The SC can indicate that the mobile's response to a short message may be sent via the same SC.
- The SC can indicate that a mobile should acknowledge receipt of a short message but discard its contents. This is useful if the SC merely needs to ascertain whether a mobile is available.
- If a mobile has been unavailable when the SC has attempted to send a short message, it can notify the SC when it is again available
- Most country-specific characters, such as the German umlaut and Greek characters, are available. This has been achieved, however, at the expense of losing CCITT IA5 control characters (with the exception of characters such as carriage return and line feed).

Network operators see SMS as a significant factor in discriminating between the services offered by a GSM phone and those of a conventional phone. Most operators see it as a value-added service for voice messaging and for applications where the transfer of textual information obviates mistranslation by voice.

13.6.4 Cell Broadcast Service

Similar to U.K. television's Teletext/Ceefax services, cell broadcast service will provide GSM mobile-phone users with information on a regular basis. GSM base stations will broadcast the information, and the mobile-phone user will be able to select the category of information to be displayed on either the phone or a computer terminal attached to the R interface. The page numbering for some of the more-common categories, such as weather reports and traffic information, has been standardized.

In all probability, the broadcasted information will stimulate a secondary action by the recipient such as calling to obtain more detailed information. Otherwise, freely broadcasting such information is difficult to justify economically. This sort of approach is commonplace in the Teletext/Ceefax services. The type of information broadcast will likely include traffic reports, accident alerts, hotel bookings, news flashes, and sports updates.

13.7 SUMMARY AND CONCLUSIONS

The services to be provided in the GSM environment are numerous and need to take into account different national requirements. The main complexity is in the interworking with the analog environment, but, despite GSM being based on ISDN technology, even the use of ISDN terminals in the mobile station will suffer from degraded performance.

It is gratifying to see the adoption of a standard for cellular data across Europe. Hopefully this will encourage the development of mobile data applications, but the implications of using data when roaming to another network are yet unknown. It must be concluded, however, that despite the sophistication in GSM technology for data, the performance envisaged for mobile access is unlikely to ever equal that for the fixed networks.

REFERENCES

CCITT Recommendations V24, V28, V25bis, International Telegraph and Telephone Consultative Committee, Geneva.

CCITT Recommendations X25, X28, International Telegraph and Telephone Consultative Committee, Geneva.

ISO 4335, Elements of Procedures, International Organization for Standards, Geneva.

ETSI GSM, Elements of Procedures, International Organization for Standards, Geneva.

    07.01, General TA functions for mobile stations, Part

    07.02, TA functions for the asynchronous services

    07.03, TA functions for synchronous services

    04.21, Rate adaptation at the MS BSS interface

    03.10, GSM PLMN connection types

    04.22, RLP

    03.43, Videotex

    03.44, Teletex

    03.40, Short-message service

    03.41, Cell broadcast

    09.xx, PLMN/fixed-network interworking

Chapter 14
Facsimile Over Cellular Radio

I. Harris
Vodata, Ltd.

14.1 INTRODUCTION

The growth in the use of Group 3 facsimile (fax) machines since the mid-1980s has been considerable, to say the least. Today, practically every business has access to at least one fax machine and, next to speech, it is fair to say that fax is the most common means of day-to-day electronic communications. So widespread is the use of fax that there is a growing tendency to use fax for nonurgent documents or even large documents that would normally have been sent by post. Despite the sophistication of data networks, fax is assured of a future well into the mid-1990s and beyond. There is now a growing desire to use fax apparatus in the cellular environment.

First, it is necessary to understand fax machines in order to understand the limitations imposed by cellular environments. CCITT Rec. T30 defines the communication protocol for Group 3 facsimile apparatus. The T30 protocol is half-duplex and comprises a binary control signal phase (BCS), to supervise the transmission of the facsimile document, and a message phase, during which the contents of the documents are sent. The document to be sent is scanned electronically and encoded as a bit stream that can be several hundred thousand bits long. This encoding and decoding is defined in CCITT Rec. T4. The BCS phase operates at 300 bps using V21 modem technology. The message phase operates at either 9,600, 7,200, 4,800, or 2,400 bps using V29/V27ter technology, depending on either machine capability or line quality. A typical page of A4 text can take 30 seconds to send at 9,600 bps. Any bit errors in transmission manifest themselves as black or white lines across the width of the received fax document, or even missing or extended lines, which can make the text unreadable.

CCITT T30 also defines an optional error correction mode. In this case, the fax document is effectively split up into blocks known as *partial pages*. Each partial page comprises 64 or 256 frames. At the end of each partial page, the receiving apparatus detects any frames in error by checking their CRC and requests their retransmission. This process can be repeated up to three times for the same partial page because some retransmitted frames may themselves be in error. The error correction mode is therefore a simple ARQ protocol. This protocol is very effective in PSTN environments where error rates on the order of 1×10^{-5} tend to become significant when transmitting the vast number of bits for a fax document. A page of A4 text may contain 40,000 bits. The error correction mode only works, of course, if both facsimile machines have the capability.

Once a call has been established between two fax machines, they exchange their capabilities (e.g., message transmission speed). The fax machines then turn on their message speed modems (normally 9,600 bps), which must first train and synchronize. The T30 protocol is then responsible for sending a sequence of bits, known as the *training check flag* (TCF) sequence, which allows the receiving fax machine to assess the line quality at that speed, and if acceptable, the document transmission begins. If unacceptable, then the fax machines will negotiate a lower message speed and recheck line quality.

14.2 FACSIMILE IN THE U.K. TACS NETWORK

Fax machines may be attached to a cellular phone using the same adapter boxes that are used to connect conventional PSTN modems. Unfortunately, fax machines contain conventional modems, which are intolerant of sustained carrier loss of more than a second or so during the message phase, and this often results in call disconnect. The modems operate in the half-duplex mode, which makes them unpredictable in a fading environment, for the reasons explained in Chapter 12. Manufacturers of portable fax apparatus intended for the mobile environment tend to provide extended carrier loss detection of several seconds, but this is only helpful if the apparatus with the extended carrier loss is receiving the image, not sending it.

Any momentary fading that affects the TCF sequence will cause most fax apparatus to retrain immediately at a lower speed. Some fax equipment can retry the TCF sequence at the same speed before changing to a lower speed.

The transmission of fax images at speeds of 4,800 and 2,400 bps result in long duration of air time and hence increased call charges. The longer the duration of transmission generally, the greater the risk of call failure, due to the inability to recover from the high number of errors in the T30 error correction mode, or alternatively due to the increased risk of carrier loss.

There is thus a tradeoff in call duration and call costs in attempting to operate at 9,600 bps, and falling back to a lower speed, compared to starting off at a lower speed and succeeding. Unfortunately, the unpredictable performance of cellular radio could result in an initial training attempt at a lower speed failing and falling back to an even lower speed; not all fax manufacturers give the user the option of starting at a lower speed.

The effect of prolonged carrier loss in the BCS phase is different from that in the message phase. The BCS phase is protected by retransmission of commands every 3 seconds for a total of three attempts. Certain BCS phase commands try for 35 seconds: for example, the negotiation which takes place at the start of a call in which the fax machines exchange their capabilities. The BCS phase is generally quite tolerant of the cellular radio environment and can sustain carrier losses of up to 6 to 8 seconds. However, under severe radio conditions, the BCS retry mechanism may fail, resulting in call disconnect. Any small burst of noise or fading during transmission of a BCS command is very likely to cause a retry.

Carrier loss in the message phase for typically 1 or 2 seconds will, in general, cause fax apparatus to disconnect from the line. Such losses are not uncommon in cellular radio, due to shadowing. The message phase is particularly sensitive to noise at the end of each page in both the error correction and non-error correction modes. Carrier loss here of less than 20 ms may result in protocol failure and possible disconnection from the line.

In order to improve the performance of T30 protocols over cellular radio, the obvious answer would seem to be to protect T30 with a forward error corrected layer 2 protocol whose performance is guaranteed. Unfortunately, the use of one layer 2 protocol to protect T30, which in itself is a layer 2 protocol, would inevitably result in a conflict of timers and retransmission mechanisms. The use of a layer 2 protocol would necessitate disregarding the T30 protocol and merely transferring the T4 encoded fax image data. This would necessitate regenerating the T30 protocol at the receiver in order to interwork with standard apparatus in the PSTN. Such a conversion would have to take place as a network service and would have to function in real time. The likelihood of maintaining end-to-end T30 procedures between the mobile fax apparatus and that in the PSTN is extremely unlikely. An alternative is to provide a store-and-forward facility as a network service. To be of benefit, such a service would require notification to the receiving party or forced delivery, both of which create operational problems in the PSTN environment.

14.2.1 Facsimile Trials for Fax Over Cellular

During 1990, Vodafone conducted extensive trials to assess the performance of fax apparatus in the U.K. TACS environment. The objectives of the tests were as follows.

- To establish the performance of standard Group 3 apparatus in a cellular environment using standard options (e.g., transmission rates);
- To identify the problems in using standard fax Group 3 apparatus in a cellular environment;
- To assess the most likely way in which standard fax Group 3 apparatus could achieve the best possible performance in a cellular environment;
- To propose a possible choice of options within CCITT/T30 that could improve performance;

- To propose ways in which T4/T30 could be modified to further improve performance.

The trials entailed sending faxes to and from a number of Group 3 fax machines attached to a cellular phone in the mobile environment. At the fixed station end, a fax machine was connected to the PSTN and its performance monitored with a facsimile protocol analyzer. The results of those trials are summarized in Section 14.2.2.

14.2.2 Error Correction vs. Non–Error Correction

The use of CCITT T30 error correction mode virtually ensures that fax images will be received error free. However, this is achieved at the cost of up to twice the transmission time without error correction, assuming a noisy environment. In a noise-free environment, the transmission time overhead with error correction is typically 15%. Any fading or carrier loss in the error correction mode will cause that part of the image to be retransmitted. In the non–error correction mode, a corrupted image will result. When the error correction mode fails, it may produce a part of the document that is perfect, whereas the non-error correction mode will produce a complete document with errors, which may or may not affect the acceptability of the quality of the document by the recipient. For example, corruptions in a numerical user data stream are obviously more catastrophic than corruption in text.

In order to assess performance of the non–error correction mode, it was necessary to define an acceptable fax image. This is defined here as one in which the number of bits received are within ±0.5% of the original. Generally speaking, any fax image outside these limits usually has at least one line that is barely decipherable by eye. (Errors in a white area of the image may not be visible, of course.)

It was found that in a mobile environment, 35% of calls produced images of acceptable quality in the non–error correction mode, compared to 73% when using the error correction mode. The error correction mode fax took about twice as long to send because of the effect of ARQ. Only fax images that were successfully completed insofar as the BCS phase was concerned were used in the analysis. In contrast, in the static cellular environment (i.e., the mobile not actually moving), the success rate using the error correction mode was 89%.

14.2.3 Error Correction Mode ARQ Performance

The error correction mode of T30 has four partial page request (PPR) attempts to recover erroneous frames. In a fading environment, only 4% of 64 octet frames needed a second PPR, compared to 18% for 256 octet frames. Unfortunately, given an error-free environment, 64 octet frames increase the overall document transmission time by some 15%. Manufacturers of fax apparatus do not often give the user a choice, and even if they did it would be difficult for the user to know which to choose. Usually, 256 octet frames are preset by the manufacturer.

The main features governing the use of fax in the analog cellular environment are summarized here.

- T30 error correction mode should be used in the cellular environment.
- Fax transmissions in a static cellular environment stand a better chance of success than in a mobile environment.
- Fax machines with extended carrier loss time-out in the message phase are less prone to premature call termination.
- Fax machines that can retrain at the failed speed can avoid unnecessary training down to a lower speed.

The results of the facsimile trials have been made available to the British Industry Facsimile Consultative Committee (BFICC), and it is possible that manufacturers will be encouraged to make their machines more suitable for a cellular radio environment.

14.3 FACSIMILE IN THE GSM DIGITAL CELLULAR NETWORK

GSM makes provision for the support of Group 3 fax apparatus through the use of the transparent and nontransparent services. The service is known as the *facsimile teleservice*.

Facsimile group 3 machines usually contain an integral modem. This needs a PSTN 2-wire analog interface at the mobile. A number of fax machines are fitted with digital (V24) interfaces. Currently within CCITT, there is no standardization of facsimile protocols across such an interface, and therefore any such protocol is proprietary. In consequence, it is not possible to connect a conventional fax machine directly to the R (or S) interface at the mobile unless it has an integral GSM fax adapter. Such a machine is envisaged by GSM, but as yet there is little indication that fax machine manufacturers are willing to integrate this functionality. Figure 14.1 shows the possible ways in which a mobile can be interfaced with a fax machine.

Figure 14.2 shows the functions required by the fax adapter. This is fundamentally a fax machine without a printer or a scanner. It has to terminate and generate the analog signals of the fax machine and terminate part of the T30 protocol that generates modem training, the T30 TCFs and certain BCS phase signals. It must also provide "ring" current to the fax machine for calls received by the mobile and be capable of converting DTMF or loop disconnect dialing to V25bis procedures for auto calling functions. A similar fax adapter function is necessary in the IWF.

As previously mentioned, the fax T30 protocol was designed to function in a PSTN environment where delays and errors are virtually nonexistent. In a GSM environment, there are inherent round-trip delays of several hundred milliseconds for the transparent service and possibly many seconds for the nontransparent service. Such delays can adversely affect the operation of the fax apparatus.

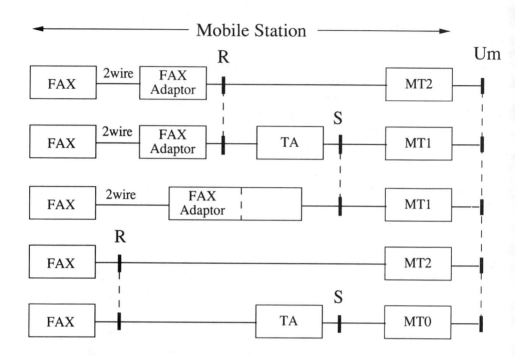

Figure 14.1 Possible mobile station configurations.

14.3.1 Transparent Facsimile Service

The philosophy of the transparent fax service is to allow the T30 protocol to pass transparently between the mobile fax adapter and the fax adaptation function in the IWF. Because of the complex nature of the GSM recommendation and the depth of knowledge required concerning T30, it is only possible here to outline the fundamental considerations necessary for the provision of this service.

T30 BCS frames are sent at the same speed as the fax message across the PLMN. The fax adapter multiplexes and demultiplexes the 300-bps BCS rate for 2,400, 4,800, or 9,600-bps message speed.

Although T30 is a half-duplex protocol, the fax adapters implement a full-duplex procedure across the PLMN. The purpose of this procedure is to convey the status of the T30 BCS protocol phases and carrier detect status of the analog fax modems. While BCS frames or message data are being sent in one direction, the status of the distant fax adapter is conveyed in the opposite direction.

The TCF sequence used by T30 to assess line quality for a given image speed is conveyed unprotected across the PLMN (except for the inherent FEC and interleaving of the transparent service), and therefore may be affected by errors in the radio environment.

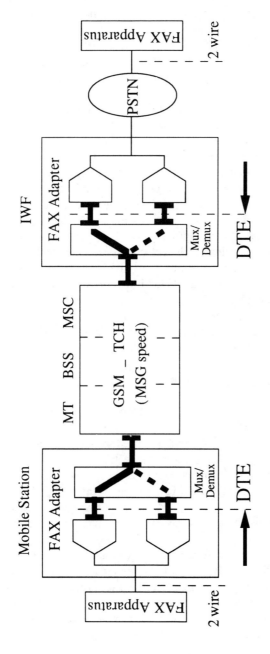

Figure 14.2 Fax adapter functions.

The TCF will therefore give an indication of the overall quality of the link, which may be impaired momentarily by a transient condition on the cellular radio path, causing the modems to train to a lower speed unnecessarily.

The operating speed of 7,200 bps is not a GSM access data rate, and therefore provision has to be made to support this rate by padding the 9,600-bps access.

The T30 error correction mode is fully supported and is of course essential to ensure fax message integrity.

Procedure interrupts require manual intervention at the fax apparatus to establish a voice call during a fax call. In the GSM mobile, there are some cumbersome technical and operational problems. First, the handset on the fax machine normally has to be lifted to complete the acknowledgment of the interrupt procedure and to speak to the other user. However, since speech cannot be sent across the R interface, the mobile handset must be taken off-hook as well to make voice conversation possible. Second, lifting or replacing the handsets in the wrong order may result in premature call clearing.

T30 nonstandard facilities (NSF) normally only work between fax machines of the same manufacturer because of the proprietary way in which they are used. Manufacturers are generally unwilling to disclose technical details concerning their use of NSF. GSM can therefore only support NSF insofar as its use of NSF does not require specific interaction with the functionality of the GSM infrastructure, such as by requiring special modulation or coding techniques. Provision is made in the mobile fax adapter to disable NSF. If this were not done, a call established between two fax machines of the same manufacturer would fail if both machines insisted on using an NSF function that required support in the GSM infrastructure.

14.3.2 Nontransparent Facsimile Service

The use of RLP to transport and protect T30 protocol elements across the PLMN requires special considerations primarily because of the potential delays that may occur. Also, because of the risk of ARQ in RLP, there is a need to buffer pages of fax message data. HDLC flags are used to keep fax apparatus "alive" in the event of fax message data not being available. Clearly, fax apparatus cannot be kept alive indefinitely and there is a finite limit to the buffer capacity in the fax adapter or the IWF.

Fax message "fill" bits used by T4 are also stripped out in order to conserve bandwidth before transmission across the radio path and are reinserted at the receiving fax adapter or in the IWF.

In the T30 error correction mode, the fax adapters make use of the Receive-Not-Ready (RNR) and Receive-Ready (RR) BCS commands to flow-control fax apparatus.

An assessment of the quality of the PLMN link is made independently of that on the PSTN through the use of TCF. If either is assessed to be of unacceptable quality, then the TCF is failed by the fax adapter in the mobile or in the IWF.

Transient impairments on the cellular radio path can result in the fax modems training to a lower speed prematurely.

T30 can encode fax message data either one-dimensionally or two-dimensionally. Two-dimensionally is more efficient, and in order to minimize delays, fax message data are always sent across the PLMN encoded two-dimensionally, irrespective of the encoding adopted by the fax apparatus.

The BCS phase has 3-second command/response timers with three attempts for most commands. Because RLP may well result in ARQ delays in excess of three seconds, then any second command attempt is absorbed by the fax adapter. This unfortunately reduces the resilience of the BCS phase commands to errors.

BCS frames are carried in a single RLP frame, which avoids segmentation delays. BCS preamble is not sent across the PLMN and has therefore to be recreated by the receiving fax adapter function before onward transmission of a received BCS frame.

14.4 SUMMARY AND CONCLUSIONS

Fax machines in the TACS environment perform as well as can reasonably be expected given the high BER that can exist from time to time, but the performance can be erratic.

The performance of fax machines in the GSM environment is unknown, but given the underlying FEC and interleaving of the transparent service, their performance should be similar to that in the PSTN, provided the transmission delays in the transparent service do not disrupt or weaken the T30 protocol. The acceptability of the nontransparent fax service seems in some doubt because of the risk of delays causing buffer overrun or premature call disconnection due to T30 timers expiring.

A more fundamental problem facing the provision of fax in GSM is the lack of availability of a mobile fax adapter or a GSM fax machine. As yet there is little indication that the fax machine manufacturers are taking fax on GSM seriously, the main reason being that no firm figures are available concerning demand.

Group 4 facsimile, which, on the surface, is more suited to the digital GSM environment, has at present a questionable future in the fixed network. The slow penetration of group 4 facsimile apparatus into the fixed network does not signal any encouraging signs concerning its use within GSM. At the time of writing, Group 4 facsimile support within GSM is an item for study only and still needs to find favor. Attention is instead turning towards the use of file transfer protocols for fax-encoded data, and this will obviate the need for complex fax adapters. Such protocols, of course, do not permit direct real-time interworking with Group 3 apparatus in the PSTN.

REFERENCES

CCITT Recommendation T4.
CCITT Recommendation T30.
ETSI GSM Recommendation 03.45, Transparent Fax Service.
ETSI GSM Recommendation 03.46, Nontransparent Fax Service.

Part V
The Future

Chapter 15
A View of the Future

W. H. W. Tuttlebee
Roke Manor Research

15.1 INTRODUCTION

Previous chapters have described the principles of cellular radio, the current and emerging analog and digital systems, and recent data applications All these systems were simply concepts, if that, at the start of the 1980s. In the 1990s cellular radio has become one of the most dynamic parts of both business and technology sectors. By all forecasts, both technological and market, this dramatic pace of progress will continue into the new millennium, leading to the implementation of new concepts of personal communication.

15.1.1 Traditional Technology Demarcations

Within Europe, European Economic Community policies aimed at developing the Single Market, since the signing of the Single European Act in 1985, have forced the pace of telecommunications development, not least in the mobile communications field [1,2]. Major new standards within the personal communications sector were initiated during the 1980s within the "traditional" technology-determined market demarcations of cellular radio, cordless telephony, and paging. Thus, at the start of the 1990s, new pan-European digital standards emerged addressing all three of these key civil telecommunications radio markets. The progress of these developments is shown in Figure 15.1. While these demarcations exist today, not only in Europe but globally, their interrelationship will change and distinctions will blur in the coming decade as the market moves from a technology focus to a service focus.

The first of these, cellular radio, comprehensively described in earlier chapters, has come from a background of providing radio telephone communications to mobile users,

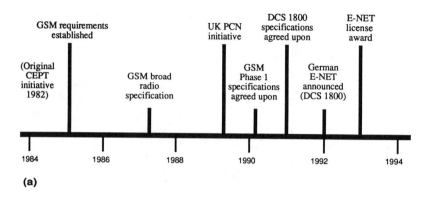

(a)

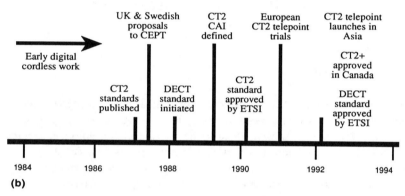

(b)

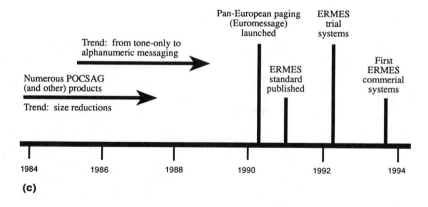

(c)

Figure 15.1 European progress to new digital standards showing traditional market demarcations: (a) digital cellular timeline; (b) digital cordless timeline; (c) paging timeline.

with high-power vehicle-mounted equipment linking with base stations at a range of several tens of kilometers. As technology has advanced, the cellular telephone terminal has shrunk to be truly portable, with low-power hand-portable telephones becoming available for the pedestrian user. Lower transmission powers imply shorter range, hence the need for smaller radio coverage cells and the advent of PCNs [3] and the DCS 1800 standard (see Chapter 8).

The complementary cordless telephony technology has sought to free the ordinary domestic telephone user from the corded link to a socket in the wall. The availability of shirt-pocket-sized digital cordless telephones to a common standard has created two completely new applications—onsite roaming using new wireless PABX equipment and cordless public access to the PSTN (telepoint). Just as pan-European standards have facilitated and encouraged the growth of the GSM and DCS 1800 systems, so also pan-European cordless telephone standards—CT2/CAI[1] and DECT[2]—are stimulating the development of these markets. Just as many new licenses have recently been awarded to operate cellular systems, so also public access telepoint systems are being widely trialed across Europe and Asia. The commercial and technology issues of the cordless telephone marketplace are comprehensively described in [4].

Paging has been perceived by some as the poor cousin of cellular and cordless, with the view that paging as a service will fade away as low-cost pocket phones become available to the general public.[3] Despite this view, the paging market has continued to show significant growth, with new equipment and facilities being introduced, including combination products, such as cordless and cellular telephones incorporating a pager— an example of boundaries beginning to blur. A pan-European standard, European Radio Messaging Service (ERMES), is at an advanced stage of definition and could potentially supersede the assorted different standards currently in use across Europe [5].

15.1.2 The Concept of Personal Communications

From the users' perspective, whether public access and onsite cordless telephony services are provided by cellular radio or by cordless technology is immaterial. Strong promotion of cellular and the newer PCN and telepoint services, together with technology advances, has served to establish the concept of a pocket telephone that may be used at home, in the office, or on the street, so that the user may both make telephone calls and be contactable wherever he or she is.[4] This is in essence of the concept of ''personal communi-

---

[1]CT2/CAI—Second-Generation Cordless Telephony Common Air Interface, ETSI standard, I-ETS 300-131.
[2]DECT—Digital European Cordless Telecommunications, ETSI standard ETS 300-175.
[3]The ''low cost'' refers to equipment and service costs.
[4]While it is true to say that this concept has been established within European industry, it will only be as the limitations of today's cordless, cellular, and PCN services become apparent, in maturing markets, that such a requirement will be mandated by the end user. At present mobility itself is a major innovation for the consumer marketplace.

cations.'' (See Figure 15.2) In the context of development of technical standards, the concept is also often referred to as *third-generation mobile systems.*

To appreciate the potential significance of personal communications, one only has to consider that major telecommunications manufacturers are currently projecting that over 40% of all telephone subscribers will be wireless by the year 2000 or shortly thereafter [6], with such wireless technologies being increasingly used for local loop applications. Personal communications is expected to develop from its niche origins in cellular and cordless to become a true commodity mass market worldwide over the next decade, as evidenced by numerous market forecasts for the new digital technology cellular and cordless products. Scale economies and increasing technology integration will support the needed equipment cost reductions. The facilities offered by mobility are such that network operators will be able to introduce a range of innovative value-added services not previously available, further stimulating market demand through new applications.

15.1.3 Chapter Structure and Content

In formulating a view of the future in this chapter, we review the early initiatives and then survey the leading-edge technology, technical standards, and market developments, highlighting key factors that will shape the development of these future systems. We also consider the role that handheld terminals for satellite communications, a new factor in the 1990s, may play in this evolution.

The viewpoint presented is essentially a European one, although the situation in North America, where recent technological progress has been rapid, is described. The reason for this stems in part from the European vantage point of the author, but also from the different approaches to technical standards in Europe and the United States. In Europe, standards are seen as a facilitator of development, encouraged within a strong framework, whereas in the United States a more *laissez faire* view leads to a less structured path of development. Thus, Europe is actively pushing to define and develop third-generation mobile systems, whereas in the United States the focus has been on shorter term competition to develop de facto technical standards.

15.2 EARLY INITIATIVES

By the late 1980s, cellular radio was clearly established in Europe as a major growth telecommunications service. At that time, the term *personal communications* was beginning to be used in forward-looking research papers presented at technical conferences, although at the time it lacked a meaning that was commonly agreed on. The term was cast into sharp relief when in 1989 the U.K. Department of Trade and Industry launched its *Phones on the Move* initiative, which ultimately led to the DCS 1800 standard (see Chapter 8). This provided new momentum to the emerging concept, which has since been maintained and developed in both Europe and North America.

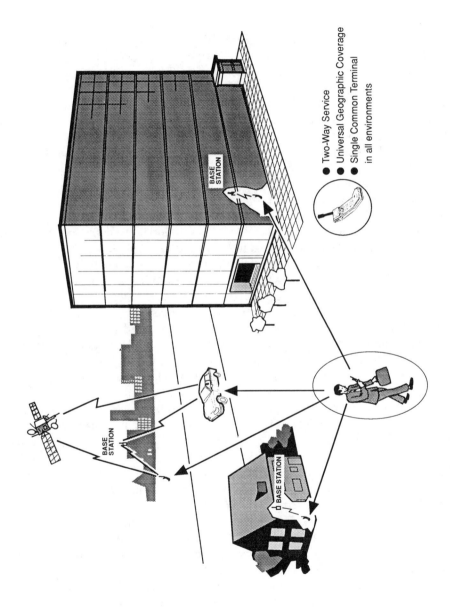

- Two-Way Service
- Universal Geographic Coverage
- Single Common Terminal in all environments

BASE STATION

BASE STATION

BASE STATION

Figure 15.2 The basic concept of personal communications.

15.2.1 RACE Program

The Research and Development of Advanced Communication Technologies in Europe (RACE) initiative was born out of a recognition of the importance of telecommunications to the future of the European economy [1]. The program comprises many interrelated projects, each of which is 50% funded by the European Community and 50% by industry. Its objectives have been both technical and cultural, insofar as it has sought to create a climate of cooperation between telecommunications administrations and industry across Europe and to facilitate a market liberalization. The technical objective of RACE has been consensus development of specific telecommunications technology and infrastructure to implement an Integrated Broadband Communications Network (IBCN) across Europe by the mid to late 1990s. Even back in the mid-1980s, mobile telecommunications was recognized as having an important place within this framework. As time has passed and the strategic importance of mobility has become increasingly recognized, this has resulted in an increasing budget share.

The first mobile activity undertaken in RACE was during the definition phase. This one-year project involved six partners and was completed in 1986. It was this project that began to actively espouse for the first time the potential convergence of cordless and cellular radio systems [7]. In hindsight, the project was visionary in highlighting the growth to be anticipated in mobile services. Its predictions that by the early 21st century, perhaps 50% of traffic carried on the public telephone network would involve mobile users, greeted with skepticism at the time, have since received increasing support. In forecasting such growth, the project concluded that mobility would have important architectural implications for the IBCN fixed-network design, and in this way paved the foundation for a mobile project within the RACE I main phase, which began in January 1988.

Under the RACE I main phase, a single project was funded addressing the specific issues of mobile telecommunications (project R1043), this time with a much larger budget and with the involvement initially of some 25 organizations—PTTs, manufacturers, and operators. The program lasted for a 4.5-year period, finishing in June 1992. The objectives of the project were to explore the potential convergence of cordless and cellular technology with a view to defining an outline of a universal mobile telecommunications system (UMTS), capable of supporting cordless and cellular-type operation from a single handset, including both radio access and fixed-network implications.

To these ends, initial work included wide-ranging research, covering such areas as fixed-network mobile functions, mobile services, radio bearer, cellular coverage, channel management, signal processing and other technologies, and mobile broadband services. The main focus of the work was on developing techniques to support a UMTS based on TDMA radio access around 2 GHz, as well as laying a foundation for a mobile broadband system using 40- or 60-GHz radio. During 1991, the focus shifted towards system definition with the formation of a permanent system definition group based in Paris and supported by other full-time system group members who would join them for monthly working meetings. Over the final 18 months of the project, an outline system concept was defined. Further work under the second phase of RACE is described below.

15.2.2 Personal Communications Networks

The U.K. *Phones on the Move* initiative in January 1989 (discussed already in Chapter 8) led to a number of bids for PCN licenses later that year. In the course of preparing these bids and subsequently, there emerged a clear, commercially driven need to understand not only the technical feasibility but also the economic viability of the concepts emerging from advanced cellular and cordless research and program such as RACE.

Without a doubt, the PCN initiative in the United Kingdom (and its subsequent endorsement by ETSI as DCS 1800) served to accelerate the pace of development of personal communications. Suddenly, leading-edge service concepts within RACE were being proposed as PCN services and propagation work at 2 GHz within and outside of the project took on new significance. The licensing of PCNs around 2 GHz to provide pocket-phone service in itself represents the adoption of concepts from the early RACE mobile research, and in Europe it represents a first step in convergence between cordless and cellular services. This in turn has stimulated the exploration of other techniques that originated in cordless telephony for incorporation within cellular systems.

15.2.3 Personal Communications Services

The strength of technical capability in civil radio telecommunications, which has existed for a long time in North America, is widely recognized and perhaps best exemplified by the comprehensive early work on cellular radio published by Bell Laboratories, which formed the basis for today's cellular radio systems [8]. During the mid-1980s, the divestiture and consequent developments in the United States were clearly a major preoccupation of the industry. Thus it was that around 1989 there was a recognition in the United States that Europe had developed a commanding lead in the commercialization of personal communications technologies, with the development of the GSM, CT2, and DECT standards and systems.[5]

Faced with the prospect of the growing commercial strength of Europe with an apparent technological lead, one response was the issue in 1990 by the U.S. FCC of a Notice of Inquiry with respect to PCSs, inviting the views of manufacturers and others on how such services should be provided. Responses to this covered a range of systems already under development, from CT2 to CDMA cellular technology; in all, over a hundred replies were filed. Since that time, numerous experimental licenses have been issued by

[5]To illustrate one view of the time, the U.S. Telecommunications Industry Association Microcell Technology Symposium held in Washington in October 1989 was publicized as "a unique experience for U.S. industrialists to explore how and why Europe has left us in the dust in developing and deploying a major telecommunications concept"!

the FCC to PCS experimenters, encouraged by a new incentive called the *pioneer's preference,* seen as an opportunity to secure commercial advantage.[6]

The digital AMPS standard, IS54 [9], described in Chapter 9, was originally conceived to provide compatible cellular coverage across the United States. One effect of these actions of the FCC, however, was to stimulate the introduction of a host of new incompatible systems. Offering potentially greater revenue from the limited spectrum, many such systems have understandably been undergoing evaluation by cellular operators. As results from such trials are becoming available, so significant new technological contributions are emerging that will influence future systems evolution.

Of all the experimental systems undergoing evaluation, the direct sequence CDMA (DS-CDMA) approach has been recognized as one of the most promising. In particular, in February 1990 a specific DS-CDMA cellular system was proposed by the Qualcomm company with claims of dramatic improvements in capacity compared to the existing AMPS standard, on the order of 18 times [10].

A successful large-scale trial of the Qualcomm CDMA system, involving five base stations and 70 mobiles, took place in San Diego in November 1991. Following a detailed report of the trial, initial technical skepticism appears to have been overcome, with the decision of the U.S. CTIA in June 1992 to support the development of a CDMA industry standard alongside the already supported TDMA digital AMPS standard. In addition to Qualcomm, numerous other manufacturers, particularly in the United States, are also trialing and developing spread-spectrum systems for cellular or indoor cordless applications. The key technical concepts of DS-CDMA cellular are described in Section 15.3.2.

15.3 TECHNOLOGY DEVELOPMENT

The fundamental technology to support personal communications essentially comprises the fixed network infrastructure elements and architecture required to locate and route a call to or from a mobile user and the radio link required to connect the user's pocket phone into the network via a radio base station. In this section, we discuss these two technology areas and current research programs supporting them. We also briefly discuss telephone terminal technology.

Coordination of research into radio and systems aspects of third-generation systems in Europe has been the objective of the COST 231 project.[7] The COST 231 group began its work in 1989 and has comprised three working groups addressing systems issues, propagation, and broadband (millimetric) aspects of future personal communications. Originally planned as a four-year program, the COST 231 activity is now expected to continue until 1996.

---

[6]The pioneer's preference offers exclusive use of reallocated radio spectrum to operators who first demonstrate new technologies or concepts for PCS.

[7]COST 231 is part of the European Cooperation in Science and Technology (COST) program coordinated from Brussels.

15.3.1 Network Technology

The late 1980s saw the development of the concept not only of mobility in the sense of the use of mobile telephones (terminal mobility), but also the concept of users being able to register their presence at a particular location to facilitate their calls being automatically routed to them at that location (personal mobility). The latter facility is implicitly needed on a mobile network to allow incoming calls, while its provision on fixed networks represents a new value-added service. In looking to the future, the distinction between fixed and mobile networks may well blur as operators seek to provide new personal and mobile services in the most cost-effective manner, maximizing reuse of existing resources. This in effect represents a service-driven requirement placed upon the network.

15.3.1.1 Intelligent Networks

The network technology development offering a cost-effective solution to such requirements is the Intelligent Network (IN). The IN concept first emerged in its current form in the mid-1980s, although many of its constituent elements predated that considerably.

The IN concept is well described in [11]. It essentially comprises the idea of facilitating separation of the intelligence from the physical switching, which, in conventional telecommunications networks, have in the past been collocated. Thus, an IN system comprises SCPs and SSPs. When an operator wishes to introduce a new service facility, it simply updates the software in its SCPs, potentially online, rather than needing to change the configuration of its switch hardware.

IN was conceived to allow flexible and efficient update of network services. While similarities exist between IN and mobile network structures, there are some differences in requirements. The mapping of IN concepts onto personal mobility has been widely and increasingly recognized, with some implementations of GSM cellular being explicitly described as IN-based, as already discussed in Chapter 6.

Although the interrelationship between IN and personal communications has yet to be fully explored, convergence is increasing. In both the IN and mobile services industry sectors, the importance of interworking between different manufacturers' solutions is recognized. Thus, for both IN and mobility, standards groups were established to develop appropriate signaling protocols as separate parts of the CCITT Signaling System No. 7— these became known as the IN Application Part (INAP) and the Mobile Services Application Part (MSAP). As these standards developed, so the functional similarities between INAP and MSAP became increasingly apparent. Given this, the potential economic benefits to the telecommunication switch manufacturers of developing a single unified signaling protocol capable of supporting both IN and mobility, rather than two separate protocols, were appreciated. Thus, more recently, the MSAP work has been subsumed into the INAP standard, ensuring that the standard will remain sufficiently flexible to support the differences demanded.

An example of such differences is information access time. Many conventional IN applications require fast initial access to service information, but are less demanding in speed of information update. By contrast, fast update is necessary for mobile applications to support handover of traffic from one base station to another as the user moves. Such demands, coupled with expected traffic levels, place stringent requirements on the signaling infrastructure and on the evolving database technologies.

The fundamental appeal of IN is its evolutionary nature; with appropriate IN architecture and hardware, IN implementation can permit a gradual evolution without the need for large capital repurchase and changeout. The network operator desiring to offer personal communications services has the flexibility to develop proprietary software features, which can offer competitive advantage without being tied to the original equipment supplier. IN architectures offer the network operator potential cost benefits and service differentiation opportunities.

15.3.1.2 Asynchronous Transfer Mode

The advent of another important network technology, broadband transmission supported by the asynchronous transfer mode (ATM), is being hastened by ongoing work within the RACE program. Compared to traditional networks, ATM networks are inherently more flexible in the way they transport signaling and user information, and may be more appropriate for supporting terminal and personal mobility. During the 1990s, many countries across Europe will introduce IN into their national PSTNs [12]. This will be followed by the introduction of ATM networks. The emerging requirements of personal communications services—for example, significantly increased signaling loads associated with user location and registration, handover between public and private networks—may be expected to influence network design as the need to integrate such requirements is increasingly appreciated.

15.3.1.3 Interconnection Techniques and Architectures

As the need for personal communications grows and smaller microcell systems are implemented, so new techniques to reduce the costs of access infrastructure are essential. Millimetric radio is a recent innovation to link base stations and mobile switching centers, thereby saving operator costs for installing fixed leased lines; further advances in this technology are anticipated, including a better understanding of propagation issues such as depolarization and scattering at 38 and 60 GHz. The transport of radio subcarriers on optical fiber is, similarly, another new technique whose technical feasibility has been demonstrated [13]. The essence of this approach is to relocate much of the complexity of a radio access base station back to a main distribution point, with the local base station simplified to contain only RF amplifiers and filters. Potentially offering economic benefits in microcell applications, variants of this approach have also been suggested to provide capacity and radio coverage improvements to existing systems (e.g., [14]).

15.3.1.4 Future Developments

January 1992 saw the start of the RACE II program, which includes research into future mobile networks and their interaction with the fixed network. The main activity in network technology is within the Mobile Networks (MONET) project, which involves the major European PTTs and telecommunications manufacturers. The project will address many of the issues addressed above, providing Europe with a strong technical input to global standards-making bodies.

15.3.2 Radio Technology

All cellular systems require radio techniques to accommodate multiple access, as discussed in Chapter 1. Early analog systems used FDMA, while the later digital systems have used TDMA combined with FDMA. CDMA had its origins in the spread-spectrum systems, which began to be developed in the 1950s, initially for military applications [15]. The advent of satellite communications saw an expansion of the role of CDMA, but suggestions of its use for civil radio applications were widely discounted until the late 1980s. At that time, they were encouraged by the so-called FCC Part 15 ruling, which permitted low-power spread-spectrum systems in the 900-MHz band to operate on a largely unregulated basis. The advent of the Qualcomm proposals in particular brought new attention to CDMA spread-spectrum as a multiple-access technology for personal communications.

15.3.2.1 DS-CDMA Cellular

The basic DS-CDMA technique is illustrated in Figure 15.3. The relatively narrowband modulating signal is spread to occupy a much wider transmitted bandwidth by multiplication with a noise-like high-rate pseudorandom code sequence (hence the term *direct sequence*). The low level noise-like appearance of the resultant transmitted signal is such that it is unlikely to interfere with other spectrum users. The signal is demodulated by correlating the received broadband signal with an identical pseudorandom sequence. This collapses the signal back to its original bandwidth and also spreads any narrowband radio signals present within the occupied spectrum, so that they now appear noise-like to the receiver. By using many different pseudorandom code sequences, multiple users may be accommodated within the same spectrum.

The key issue, which for many years had been thought to preclude the use of CDMA for many applications, was the so-called "near-far problem." If a narrowband interfering signal falls within the spread bandwidth and its amplitude is significantly greater than the

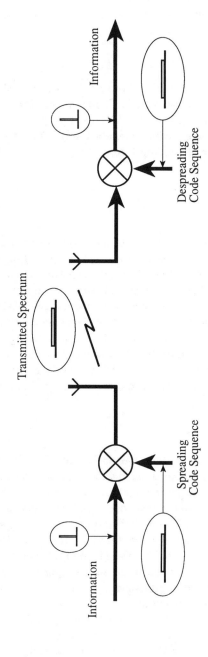

Figure 15.3 Basic principles of DS-CDMA spread spectrum.

wanted (spread) signal,[8] then the interferer would still remain sufficiently strong after despreading to prevent correct demodulation of the wanted signal. Further, for CDMA to operate successfully as a multiple-access technique, as required for cellular radio, all the accessing signals must be received at a similar amplitude. The emerging solution to this problem has been the application of power control, whereby the cellular base station controls, in real time, the transmission power of the users' mobile terminals. In GSM, power control is a feature that has been added to increase system capacity by reducing average interference levels; for CDMA, however, power control is essential to the successful operation of the basic concept.

The application of power-controlled CDMA to cellular radio enables the same frequency spectrum to be reused in all the cells, in effect rewriting the rules for frequency reuse described in Chapter 1. The application of dynamic power control and common frequency usage between cells, with different mobiles using different codes, spreads the effects of other unwanted user signals, an effect known as *interferer diversity*. It also gives rise to a new approach to handover, generally referred to as *soft handoff,* whereby, at any one moment, a mobile user may be served by more than one radio base station. The impact of these mechanisms is to increase the system capacity for a given infrastructure density, but at the expense of base station and infrastructure complexity; calls need to be simultaneously routed by the infrastructure to multiple base stations, with additional signaling needed to support this.

15.3.2.2 Advanced TDMA

It has been argued that to compare 1990s CDMA with GSM, a system specified in the mid to late 1980s, is an inappropriate comparison given the recent pace of technological advances. Significant further potential exists with advanced TDMA—indeed, it is possible that some of the techniques implicit in the new CDMA systems may also be applicable to TDMA. In terms of future TDMA systems, the rigid frame structures of GSM may be expected to give way to more flexible dynamic assignment with reservation protocols; adaptation of data rate and modulation level, or even modulation scheme, are also concepts being researched.

Within the RACE II program there exist two radio access projects, addressing CDMA and advanced TDMA, respectively. Each of these is developing new techniques and concepts and will implement and trial test-bed solutions with a view to providing technical input to the standards process. From a pure technology viewpoint, leaving aside the political factors that could shape standards evolution, it is not at present clear which radio access technology will emerge as a preferred option; indeed, a hybrid approach could well eventually be selected for a third-generation standard, combining the best features of CDMA and TDMA.

---

[8]On the order of the ratio of the spread bandwidth to the modulating signal bandwidth.

15.3.3 Pocket Phone Terminal Technology

The trend in the 1980s was to ever smaller cellular radio handsets, with the balance of new equipment sales moving strongly from vehicle-installed equipment to hand-portable phones. The advent of pocket cordless telephones reinforced this trend to small terminals in the mind of the user. These trends were made possible by developments in silicon digital technology (CMOS microprocessors, DSPs, ASICs, and PLDs), combined with developments in device packaging and board assembly (SMT and multilayer boards). The considerably increased technical complexity of the newer GSM and DAMPS phones has been largely offset by the anticipated progress in digital silicon and other implementation technologies. Despite concerns that the initial digital cell phones would be larger than their analog predecessors, hand-portable GSM telephones became available very soon after the first vehicle-mobile products.

Looking to the future, further implementation technology advances are anticipated to facilitate the truly pocketable lightweight personal phone, with increasing use of multichip modules, chip-on-chip and chip-on-board technologies. 3.3v silicon, additional metallization layers, and yet finer geometries together promise significant power consumption savings, thereby giving longer talk and standby times. Further power reductions await new lower voltage semiconductor technology. Integrated analog and mixed technologies may also be expected to facilitate a greater degree of radio circuit integration, possibly allowing the implementation of less conventional radio architectures. Further developments in SAW filtering and active filtering techniques also hold promise.

At the same time, improvements in battery technology will offer the user more choice between talk time, size, and weight parameters, as well as give the manufacturers and system developers more opportunity to use complex signal processing.

15.4 TECHNICAL STANDARDS

The development of technical standards is key to the emergence of a mass market for any technology-based product or service. In Europe, the motivation for the GSM cellular standard was one of creating a large, unified market with potential for pan-European roaming; a similar philosophy was implicit in the original thinking underlying the IS-54 standard in the United States.

Within the field of telecommunications, standards play a particularly significant role, since the very nature of telecommunications involves technical interfacing across distance and hence across national or regional boundaries. Since its founding in 1866, the ITU, now part of the United Nations, has sought to establish worldwide cooperation, with technical standards being created by its consultative committees—the CCITT and the CCIR. Coordination in Europe was traditionally the responsibility of the CEPT. The creation of ETSI in 1988, however, represented a step change in the European standards environment, facilitating an acceleration of the process of standards creation in Europe. Further details of the standards bodies and processes may be found in [4].

15.4.1 Future European Standards for Personal Communications

ETSI was formed as a result of proposals from the European Community in the 1987 Green Paper on the liberalization of the telecommunications market [1]; the background to and role of ETSI as envisaged in its early days are described in [16]. Since 1988, the responsibility for creating technical standards has passed from the CEPT to the ETSI technical committees.

The two main ETSI technical committees dealing with radio-based telecommunications have been Radio Equipment and Systems (RES) and Special Mobile Group (SMG), formerly entitled GSM. RES has developed cordless telecommunications standards (especially DECT), and SMG the GSM cellular system. In addition, an additional subcommittee of the Network Architecture (NA) technical committee is developing standards for universal personal telecommunications (UPT), of relevance to both mobile and fixed telecommunications.

In recognition of the progress of mobility research within RACE, an ETSI ad hoc UMTS group was established, which met between August 1990 and April 1991 under the auspices of RES. The report of this group recommended establishing a permanent subcommittee to establish European standards for UMTS. The ETSI Technical Assembly duly authorized such an activity, transferring responsibility, however, to SMG. A preparatory meeting of this new subcommittee, SMG5, in November 1991 was followed by a formal inaugural meeting in March 1992. These initial meetings largely addressed procedural matters, methods of working, time scales, and limited discussion of technical requirements, while subsequent meetings have begun to address detailed technical issues. As other SMG activities on the GSM standard are reduced, so other SMG subcommittees will increasingly participate in the definition of UMTS.

It is planned that basic technical principles should be agreed on by 1994-95 and detailed protocols by 1996-97 to enable service to begin shortly after the year 2000. Technical results from the RACE II program, as well as from independently funded company research, are expected to form the main sources of technical input to the SMG debates to achieve these time scales.

15.4.2 Future World Standards for Personal Communications

Work on third-generation mobile systems within CCIR predated that of ETSI by several years, initially being undertaken by the Interim Working Party 8/13 (IWP8/13) of Study Group 8 of the CCIR, established in 1986. Following the May 1990 plenary of the CCIR, this group was re-established as Task Group 8/1 (TG8/1), reflecting a recognition of the increasing importance of its work. It has subsequently met in May 1991, January 1992, and October 1992.

TG8/1 has produced a range of documentation (available from ITU, Geneva) detailing their progress to date under the generic title "Future Public Land Mobile Telecommunications Systems'' (FPLMTS). Separate documents, at different stages of drafting, address the following topics.

- Overall concept;
- General system description;
- Service requirements;
- Network architectures;
- Speech coding requirements;
- Quality of service;
- Satellite interworking requirements;
- Network interfaces;
- Radio interfaces;
- Network management;
- Application in developing countries.

A number of different architectural scenarios for the provision of personal services are described by TG8/1 from a perspective of not wishing to exclude any technical possibilities; these have included various approaches to infrastructure, ranging from separate from to integral with the PSTN. Of note is the inclusion of a satellite component in the FPLMTS concept.

Reflecting the increasing integration between the fixed-network and radio aspects of mobile telecommunications, the CCITT is also contributing to the work of TG8/1. As well as sending *rapporteurs* to several relevant CCITT committees, joint meetings between the CCITT and CCIR have also been held (e.g., with CCITT SG1 in autumn 1991). While this is very positive from the viewpoint of technology integration, it is not without the obvious problems arising from lack of familiarity of CCITT personnel with FPLMTS concepts and terminology.

Much of the work of TG8/1 during 1990-91 was focused on the World Administrative Radio Conference (WARC) held in Spain in February 1992, at which spectrum allocation for FPLMTS was one of the high-profile debates. Initial post-WARC work focused on achieving the planned schedules for standards development.

15.4.3 Radio Spectrum Allocation

The 1992 WARC of the ITU was of major significance in determining the future commercial growth possibilities of cellular and personal communications worldwide, since such systems fundamentally require the allocation of radio spectrum. Radio spectrum is of course a scarce resource and satellite and other potential services were competing at the WARC for the same spectrum.

The conference resulted in the elevation of the status of land mobile services to joint primary status, coequal with fixed services, in the band 1,700 to 2,450 MHz. With respect to third-generation systems, a total of 230 MHz was designated on a worldwide basis for the implementation of the terrestrial component of FPLMTS from the year 2000, specifically the bands 1,885 to 2,025 MHz and 2,110 to 2,200 MHz.[9] Further, the subbands 1,980 to 2,010 MHz and 2,170 to 2200 MHz were designated to support the satellite component of FPLMTS from the year 2010 onwards. The allocations of WARC92 are summarized in Figure 15.4.

These FPLMTS allocations were made based on spectrum requirement estimates emerging from both RACE and CCIR and are consistent with envisaged ETSI and CCIR time schedules. The success of the WARC in this respect cleared the way for the development of third-generation systems to support personal communications.

15.4.4 Interrelationship of UMTS and FPLMTS

The interrelationship of ETSI SMG5 and CCIR TG8/1 continues to evolve. It remains to be seen how closely the two committees will work together in the next few years in developing future standards. Moves towards alignment of the CCIR and ETSI schedules have been one of the encouraging post-WARC developments. Further, there exists significant functional overlap in terms of people between the two committees, which should at a pragmatic level also encourage the process of cooperation.

It is not clear how the detailed ETSI UMTS and CCIR FPLMTS technical standards will converge. Resolution COM4/4 from WARC92 states that any implementations of FPLMTS should conform to CCIR and CCITT recommendations. This means that the ETSI SMG specifications should conform to those emerging from TG8/1. However, it is still possible that the ETSI standard could form a more tightly specified subset of a CCIR generic standard. While it may be clearly desirable at one level to see a convergence of technical standards, other considerations will come into play.

15.5 MARKET ISSUES

In attempting to forecast the future development of personal communications, the role of economic and market factors cannot be overstated. Much of the impetus for the digital radio telecommunications technology developments of the 1980s in Europe and North America stemmed from the rapid growth of the analog cellular radio markets. Likewise, the pace of future progress will reflect the economic climate of the 1990s.

The marketplace success of the new digital cellular and cordless technologies will be a key factor. If takeup of these products falls significantly behind real expectations, then manufacturers' investments in follow-on technology will be reduced. Predictions for both cellular and cordless markets during the 1990s are explosive, largely based on extrapolations of earlier performance and anticipated price reductions facilitated by economies of scale. Many observers would adopt more cautious expectations; even conservative

---

[9]The time gap between the WARC and the allocation coming into effect is required to allow existing users to vacate the spectrum without jeopardizing existing investments in equipment.

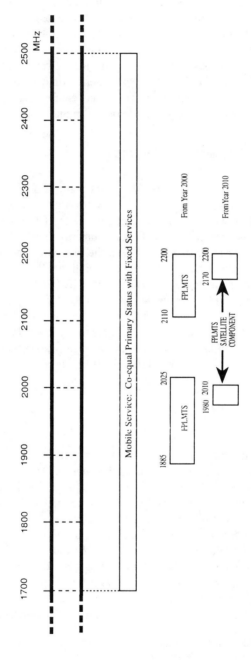

Figure 15.4 Spectrum allocations for FPLMTS.

forecasts, however, still indicate enormous potential. Ongoing liberalization in Europe can be expected to result in more new mobile communications operators being licensed; likewise, the use of mobile technology in the Eastern European states will contribute to market growth.

Future developments will also be influenced by the success of other related telecommunications technologies. Recent forecasts, for example, have suggested that the early 1990s will see the commercial success and proliferation at long last of videotelephony. If such forecasts do prove correct, it remains to be seen what new requirements or applications for mobile videotelephony may emerge. Similar arguments may be applied to the interaction of personal and mobile communications with the other new technologies of notebook and palmtop computers, multimedia, and virtual reality.

The trend of mobile and personal communications from being technology-led niche markets to customer-driven mass markets will stimulate new service innovations. The interlinking of cellular, cordless, and paging services is one possible scenario that is already emerging, with cellular and cordless handsets with integrated pagers already becoming available. The advent of personal numbering and its availability in conjunction with mobile networks will also further the integration between mobile and fixed networks and hasten the development of unified personal communications. In this respect, the pace and success of implementation of IN will also influence the advent of new personalized services on fixed, mobile, and integrated networks.

15.6 THE ROLE OF SATELLITE SYSTEMS

The original concept that satellite systems had a role to play in personal communications stemmed from several sources. Within the CCIR IWP8/13 group, satellite communications was perceived to offer a potential mechanism for direct service provision to users in those less-developed countries lacking a strong telecommunications infrastructure, and thus was incorporated into the concept of FPLMTS at an early stage; studies of the future role of satellite communications were also being undertaken within a sister CCIR committee, IWP 8/14. Independently, studies of land mobile satellite systems (LMSS) were being undertaken by the European Space Agency (ESA) based on the similar concept of satellite coverage for the less developed regions of Europe, for which it would be economically inappropriate to provide conventional GSM coverage.

As with cellular and cordless, technological progress has facilitated size and cost reductions of satellite communications equipment over recent years, which has led to the commercial development of new mobile services and equipment. The commercial launch of the Inmarsat-M system in 1992 is a step in this direction, offering communications-quality voice, group calls, fax, and 2.4-Kbps data services to lightweight, low-cost terminals installed in vehicles or in briefcase-sized personal equipment. Based on geostationary satellites, it does not provide handheld terminal service, which would require a larger effective radiated power (transmit power/antenna size combination).

In North America, steps towards spectrum allocation for mobile satellite service have stimulated the beginnings of convergence between satellite and cellular services, with preliminary agreements in early 1992 between American Mobile Satellite Corporation[10] (AMSC) and several cellular telephone manufacturers to pursue the supply of AMSC satellite/cellular dual-mode mobile telephones. AMSC's system would allow cellular subscribers to communicate via the AMSC system when outside of a cellular coverage area.

Despite such agreements, the integration of LMSS with conventional cellular systems has been the subject of only limited study to date, with no clear consensus yet emerging. While integration in terms of full technical harmonization would appear impractical on a short time scale, a useful conceptual model of different levels of integration has been outlined [17], which proposes four increasing levels of integration, each of which includes the basic concepts of the previous ones.

At the lowest level of integration—*geographical integration*—the land mobile and satellite systems simply provide overlapping coverage, but are based on independently defined techniques and offer independently defined services; this is in effect today's situation. At the second level—*service integration*—the satellite component is envisaged as supporting a subset of the land mobile services, albeit perhaps at a degraded quality of service, given technical constraints. At the next level—*network integration*—a fixed user would be able to call a mobile user without having to specify call routing information; that is, a single calling number would suffice to identify the called subscriber, whether accessed by satellite or land mobile access. The final level of integration would be full *system integration,* whereby the satellite system is fully integrated with the land-based coverage.

The task of coordinating work on the integration of space and terrestrial mobile networks has been assigned within Europe to COST 227 (similar in organization and working procedures to COST 231). The objective of COST 227 is to study the feasibility of systems for mobile communications consisting of two integrated subsystems—satellite and land based. Such integration should maximize commonality of services, commonality of mobile terminal functions, and commonality of fixed infrastructure facilities. The COST 227 work began in 1991 and is due to be completed in 1994, although its life may be extended.

15.6.1 Iridium

While satellite personal communications has been proposed for some years, the development that brought it to widespread public prominence was the announcement in June 1990 by Motorola of their intention to develop a worldwide, digital, satellite-based, cellular, personal communications system, to be known as *Iridium.* See Figure 15.5.

---

[10]AMSC, a consortium comprising several U.S. companies who initially all filed separate petitions with the FCC for spectrum allocation, was licensed by the FCC as a common carrier satellite service. AMSC plans to launch its first satellite in 1994. It has recently extended its initial service concepts as detailed below.

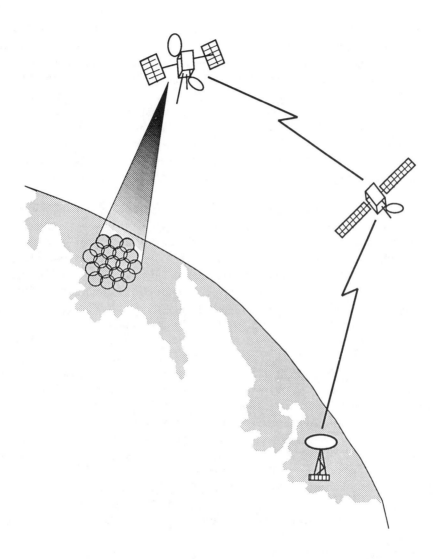

Figure 15.5 Satellite cellular concept.

The system was not intended to compete directly with conventional cellular systems, but rather to complement them. It would support fixed users in regions currently without telecommunications infrastructure, mobile users in regions currently lacking mobile services, and would provide a backup service where mobile systems already existed. The system would allow solar-powered telephones to be provided in remote villages, as well as allow international businessmen to be confident of always being available and able to contact high-level staff. Speech, paging, and data services were envisaged. Several competitor systems have also been proposed, although to date Iridium has been the most publicized and the one with the most technical detail made available in the public domain.

The Iridium proposal was originally based around a constellation of 77 small (about 320 kg) satellites; hence its name—the element iridium has an atomic number of 77, implying 77 electrons circling its central nucleus. The satellites were planned to orbit the earth in polar low-earth orbit (LEO) at an altitude of about 800 km in seven corotating planes, each plane containing 11 satellites, with each satellite circling the earth once every 100 minutes. The rationale for the choice of constellation is given in [18]. During 1992, the implementation concept was revised to allow the number of satellites to be reduced to 66.

Within the Iridium concept, satellites would be accessed by portable, mobile, or transportable terminals using low-profile antennas throughout the world. Satellites would be internetworked to provide an orbiting backbone infrastructure capable of routing calls onwards to other system users using intersatellite links or via gateway ground stations into the terrestrial PSTN. The satellite constellation would cover the earth with a network of cells, with a diameter of about 400 miles, enabling frequency reuse to be employed as with conventional cellular systems; a modified seven-cell reuse pattern is proposed. Each satellite would be capable of projecting a 37-hexagonal cell pattern; the satellite would use fixed geometry interlaced spot beams; hence, the pattern would move across the earth's surface as the satellites orbit the earth. Thus, whereas in a terrestrial cellular system the users are mobile and handover is performed as they move between cells, with Iridium the users would be stationary (relatively), with handover performed as the cells themselves move. Individual cells would be switched off to reduce cell coverage overlap at high latitudes and to restrict coverage where not required.

Iridium pocket phone radio access would operate around 1.6 GHz using QPSK modulation, a TDMA/FDMA access technique and transmission power of 0.6W, and would build upon GSM experience with similar features and protocols. It would support 4.8-Kbps vocoded speech using VSELP coding with FEC; it would also have the capability of upgrading to 2.4 Kbps as improved vocoders are developed, as well as 2.4-Kbps data. Intersatellite and gateway links would operate in the 20- to 30-GHz region. See Figure 15.6.

Dual-usage terminals are proposed such that, before connecting a call via satellite, access to a local cellular network would first be attempted. This would avoid unnecessary burdening of the system with urban-originated calls and would automatically offer the user least-cost routing. As well as a telephone handset, an additional pager unit, offering an additional 15- to 20-dB link margin, would alert users to incoming calls when in unfavorable locations, such as deep within buildings.

Since the original announcements in 1990, Motorola has refined its initial proposals. Public statements suggest that the detailed specifications should now be complete, with first satellite launch in 1994 and system operation by 1997. Major regulatory and legal

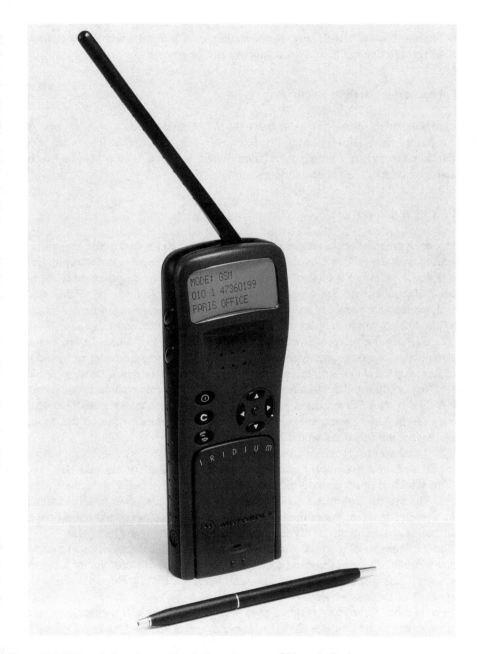

Figure 15.6 Iridium dual-mode portable telephone (courtesy of Motorola, Inc.).

hurdles remain, however, with several other companies in the United States fighting to secure the right to operate their own competitive offerings.

15.6.2 Alternative Satellite Ventures

Although Motorola's proposals have been the most widely publicized, several other U.S. companies have also applied for spectrum to offer similar satellite-based mobile communications services. Outside the United States, Inmarsat is also planning to offer satellite-based personal telephone services.

15.6.2.1 U.S. Initiatives

Ellipsat, one of the smaller companies in the field, was in fact the first to file an application for spectrum with the FCC in November 1990 with what has become known as its *Ellipso* system. Ellipsat proposes a total of 24 satellites in a range of elliptical orbits between 400 and 2,000 miles and the use of CDMA modulation. The use of CDMA would potentially permit coexistence of multiple operators fielding competing systems—although whether the market would support multiple operators is another matter!

The *Globalstar* concept was announced in October 1991 by a consortium including Loral, Qualcomm, Aerospatiale, Alenia, and Alcatel. The concept, developed by the joint venture company Loral Qualcomm Satellite Services, is also based on CDMA technology, pioneered for cellular and for satellite-based vehicle positioning applications by Qualcomm. A total of 24 LEO satellites, operating purely as repeaters, are initially envisaged to provide North American coverage, with a further 24 planned to extend this to world coverage; switching functions would be terrestrially based.

The *Tritium* concept was also launched in October 1991 by the Hughes Aircraft company. Based on geostationary satellites rather than LEOs, Tritium will also involve AMSC, of which Hughes owns 30%, and would presumably build upon their planned activities. The system anticipates the use of 4.8-Kbps vocoding with 6-kHz channeling FDMA. Vehicle, or fixed, rather than personal telephones are envisaged due to the high transmit power requirements involved with the use of geostationary satellites.

The TRW proposal, entitled *Odyssey,* would employ 12 satellites in an equatorial orbit at an altitude of several thousand miles. The *Aries* system of Constellation Communications, Inc., (CCI) would comprise 48 low-orbit satellites. Little technical detail is available on either of these schemes.

Several other contending systems have been proposed by U.S. consortia. Amongst these, a late entry to the fray was *Celsat,* offering in spring 1992 a system providing U.S. coverage based on just two geosynchronous satellites using very large multiple beam antennas to create a complex network of ground coverage cells. Celsat claim that use of such state-of-the-art technology would permit costs, and hence call pricing, to be significantly lower than any of the other currently proposed systems.

15.6.2.2 *International and European Initiatives*

At the time that Iridium was first announced, collaboration had been established between Motorola and Inmarsat, the international satellite organization comprising telecommunication operators from some 64 member states. Inmarsat, formally established in 1979, has steadily increased its range of satellite-provided telecommunications services from services to maritime users to include airborne and land mobile users. Subsequent to the initial links between Motorola and Inmarsat, the latter announced its *Project 21* team in September 1991, formed to prepare possible scenarios for a fourth-generation of Inmarsat satellites, capable of supporting handheld terminals, to be launched around 1997 [19]. To be known as the Inmarsat-P service, little technical detail of the Inmarsat proposals is yet available. The formal go-ahead for the full development of such a capability has yet to be given by the Inmarsat council at the time of writing. If this is given, however, the system would be likely to receive widespread international support, given the membership of the Inmarsat organization. In this respect, Inmarsat is perhaps the best placed of all the contenders to overcome the anticipated regulatory hurdles.

The European Space Agency concepts for LMSS have revolved around the use of transparent geostationary satellites providing European coverage using either a single beam or multiple spot beams. The former will be trialed using the *European Mobile System* (EMS) payload, expected to fly in 1993, and the latter using the *Artemis* satellite, expected to fly in 1994. Integration between the ESA LMSS concepts and GSM is envisaged at nothing deeper than at the level of service integration, although the prospect for integration at the deeper level of system integration in the next generation has been recognized and proposed to the CCIR [17].

15.6.3 Implementation Practicalities

Very few of the recent proposals for satellite-based LMSS and pocket phone services will come to fruition, simply because of the huge investments which would be needed; rather, a rationalization will occur. Financing of these new ventures will require bold and rich investors convinced of both the market potential and the technical soundness of the proposals. Estimates for the cost of Iridium vary between $2 to $3 billion, with a financial break-even at around 0.5 million users. Costs for geostationary systems could exceed this.

Regulatory issues for satellite-based telecommunications should also not be minimized, given the potential of a direct satellite link to bypass the local telecommunications operator and given the potential competitive threat to revenues of existing satellite telecommunication systems. Resale of system capacity to existing fixed-network operators and telecommunications administrations is one approach to reducing such apparent competition.

One of the major hurdles to system implementation, however, was overcome with the WARC decisions in early 1992 to allocate spectrum for such satellite services. In addition to shorter term allocations, which will facilitate development of Iridium or a similar service, the decision to designate spectrum for the satellite component of FPLMTS by the year 2010 should encourage development and convergence of such satellite and terrestrial mobile and personal communications.

Despite such hurdles, several consortia have announced intentions to proceed with experimental systems, with the FCC awarding experimental satellite licenses to Motorola, Ellipsat, and Constellation in August 1992.

15.7 CONCLUSIONS

The foregoing sections have outlined the major developments likely to shape the evolution of the cellular industry over the next decade or so. We conclude this discussion by drawing together some predictions for the future development of the industry. Any such predictions will invariably present only a partial and inaccurate picture. However, the foregoing material should enable the reader to modify these forecasts and to develop his or her own extrapolations in the light of subsequent developments as the decade progresses.

15.7.1 The Short Term—Next Steps in Cellular

In the short term, say the period 1993 to 1996, we may expect to see a gradual entry and market growth of the new digital cellular systems, GSM and DAMPS. Complementing these, we may expect to see new PCNs/PCSs also being implemented, predominantly in urban areas. The pace of such deployments, however, will be influenced by the economic climate in Europe and the United States and may not be as rapid as forecasted by some observers. The effect of multiple digital cellular standards in the United States remains to be seen; one would, however, expect it to weaken the overall U.S. position in the world market relative to Europe.

As these cellular services are deployed, so also cordless telephony services will grow, particularly in Europe and Asia. An increasing degree of interaction and the beginnings of interworking between cordless and cellular, and bundling with paging, is also to be expected within this time frame.

15.7.2 The Medium Term—Future European and World Standards

Later in the decade we may expect to see technical standards emerging for third-generation PCSs—UMTS/FPLMTS—which will offer new and enhanced services and facilities. Based on research in Europe in RACE and other forums and on research and trials of experimental systems in North America, the degree of universality or synergy that will be secured with these standards remains to be seen. The choice of radio access technique currently remains open, but it should begin to clarify by the mid-1990s; the network architecture implications should also become apparent in this time frame. If projections for the growth of user mobility are only partly correct, the implications for mainstream telecommunications will be profound—personal communications will no longer be the niche service or overlay network that cellular radio comprised during the 1980s.

Satellite-Based Personal Communications

A huge technical gulf still exists between conventional cellular systems and satellite-based mobile services. This is expected to reduce dramatically over the next few years to the point where service compatibility may be expected, and shortly thereafter, perhaps, even network compatibility may be expected. Full systems integration is undoubtedly farther off and is unlikely to be seen until 2010 or beyond; indeed, it is quite conceivable that by that time other new evolutionary forces will have emerged, which could influence the desirability or viability of such convergence. If investors can be found, satellite-based personal communications are likely to emerge as a major growth market to complement cellular in the more sparsely populated regions of the world over the next decade.

REFERENCES

[1] The European Community Green Paper on "The Development of the Common Market for Telecommunication Services and Equipments," COM(87) 290, Brussels, 1987 (See also the European Community Green Paper on "Development of European Standardisation: Action for Faster Technological Integration in Europe," COM(90) 456, Brussels, 1990).

[2] Tuttlebee, W. H. W., "Standards and Technology for Personal Communications," *Personal Communications Networks Conference*, IBC Technical Services, London, April 1991.

[3] Gardiner, J., ed., *Personal Communication Systems*, Mobile Communication Series, Artech House, May 1993.

[4] Tuttlebee, W. H. W., ed., *Cordless Telecommunications in Europe*, ISBN 3-540-19633-1, Springer-Verlag, December 1990. This reference provides a comprehensive review of the commercial and technical aspects of cordless telephony and the standards processes, as well as additional detail on future systems evolution. Recent commercial developments are summarized in "Cordless Personal Communications," *IEEE Communications Magazine*, December 1992.

[5] Lax, A., *Radiopaging*, Mobile Communications Series, Artech House, May 1993.

[6] Jansson, H., "The Future of Cellular Telephony," *XIII International Switching Symposium*, Stockholm, May 1990. (See also H. Jansson, J. Swerup, and S. Wallinder, "The Future of Cellular Telephony," *Ericsson Review*, No. 1, 1990.)

[7] Gibson, R. W., R. J. G. MacNamee, and S. K. Vadgama, "Universal Mobile Telecommunications System—a Concept," *European Seminar on Mobile Radio Communications*, Brussels, April 1987.

[8] MacDonald, V. H., "The Cellular Concept," *Bell System Technical Journal*, No. 58, January 1979, pp. 15–42.

[9] "Cellular System Dual-Mode Mobile Station—Base Station Compatibility Specification," Interim Standard IS-54, U.S. Electronics Industry Association/Telecommunications Industry Association, May 1990.

[10] Gilhousen, K. S., I. M. Jacobs, R. Pandovi, A. J. Viterbi, L. A. Weaver, Jr., and C. E. Wheatley III, "On the Capacity of a Cellular CDMA System," *IEEE Transactions on Vehicular Technology*, Vol. 40, 1991, pp. 303–312.

[11] Ambrosch, W. D., A. Maher, and B. Sasscer, *The Intelligent Network*, Springer-Verlag, 1989.

[12] Shorrock, D., ed., *European Communications—Technologies and Regulations of the Single Market,* ISBN 0-86353-182-2, Blenheim Online, 1989.

[13] Cooper, A. J., "Fibre/Radio for the Provision of Cordless/Mobile Telephony Services in the Access Network," *Electronics Letters,* Vol. 26, 1990, pp. 2054–2056.

[14] Lee, W. C. Y., "Smaller Cells for Greater Performance," *IEEE Communications Magazine,* November 1991, pp. 19–23.

[15] Scholtz, R. A., "The Origins of Spread-Spectrum Communications," *IEEE Transactions on Communications,* Vol. 30, May 1982, pp. 822–854.

[16] Gagliardi, D., "The Development of Standards," *Proceedings of the Pan-European Digital Cellular Radio Conference,* IBC Technical Services, Munich, 8, 9 February 1989.

[17] Arcidiacono, A., "Integration Between Terrestrial-Based and Satellite-Based Land Mobile Communications Systems," *Proc. Second International Mobile Satellite Conference,* Ottawa, June 1990, pp. 39–45.

[18] Leopold, R. J., "Low Earth Orbit Global Cellular Communications Network," *VSAT '90 and European Satellite Users' Conference,* Luxembourg, November 1990. (See also R. J. Leopold, "When Telephones Reach the Smallest Village," *Global Communications Magazine,* September/October 1991, pp. 28–35.

[19] Wood, P., "Mobile Satellite Services for Travellers," *IEEE Communications Magazine,* November 1991, pp. 32–35.

Acronyms

ACCH	associated control channel
ACH	access channel
ACS	adjacent channel
ACU	antenna combination unit
ADC	American digital cellular
ADM	administration center
AGC	automatic gain control
AGCH	access grant channel
AI	area identification
AMC	automobile switching center
AMPS	advanced mobile phone system
AMSC	American Mobile Satellite Corporation
ANSI	American National Standards Institute
ANT	base station antenna
ARFCN	absolute radio frequency channel number
ARQ	automatic repeat request
ASIC	application-specific integrated circuits
ATM	asynchronous transfer mode
Au	authentication
AuC	authentication center
AuR	authentication register
AXE	Ericsson telephone exchange (trade name)
BCC	base (station) color code
BCCH	broadcast control channel
BCE	base station control equipment
BCH	Bose-Chaudhuri-Hocquenghem
BCIE	bearer capability information element

BCSP	binary control signal phase
BER	bit error rate
BFI	bad frame indicator
BFICC	British Industry Facsimile Consultative Committee
BIS	busy-idle status
BS	base station
BSC	base station controller
BSC	base station switching center
BSI	base station interface
BSS	base station subsystem
B·T	bandwidth bit-rate product
BT	British Telecom
BTS	base transceiver station
CA	charge analysis
CAC	common access channel
CACH	common access channel
CC	call control
CC	country code
CCCH	common control channel
CCH	calling channel
CCH	Committee on Harmonization
CCI	Constellation Communications, Inc.,
CCIR	International Committee on Radio
CCITT	International Consultative Committee for Telegraph and Telephone
CDLC	cellular data link control
CDMA	code-division multiple access
CDR	charging data recording
CDVCC	coded digital verification color code
CELP	code excited linear prediction
CEPT	Committee of European Posts and Telecommunications
CIR	carrier-to-interference ratio
CMAC	maximum power in advanced mobile phone system
CMOS	complementary metal oxide semiconductor
COST	European Cooperation in Science and Technology
CRC	cyclic redundancy check
CTIA	Cellular Telecommunications Industry Association
CTS	clear-to-send
CU	control unit
CW	continuous wave
DAMPS	digital advanced mobile phone system

DC	data channel
DCC	digital color code
DCN	data communication network
DCCH	dedicated control channel
DDI	Daini-Denden, Inc.
DECT	digital European cordless telephone
DFE	direct feedback equalizer
DLL	data link layer
DMSU	digital main switching unit
DNIC	data networking identification code
DPSK	differential phase-shifting key
DRX	discontinuous receive
DS-CDMA	direct sequence code-division multiple access
DSAT	digital supervisory audio tone
DSP	digital signal processing
DSRA	dynamic slotted reservation aloha
DSRR	digital short-range radio
DST	digital signaling tone
DTE	data terminal equipment
DTI	Department of Trade and Industry
DTM	digital terrain map
DTMF	dual-tone multifrequency
DTX	discontinuous transmission
DVCC	digital verification color code
EAMPS	extended advanced mobile phone system
EFTPOS	electronic point-of-sale finance terminal
EIA	Electronic Industries Association
EIR	equipment identity register
ERMES	European Radio Messaging Service
ERP	effective radiated power
ERS	Ericsson Radio Systems
ESA	European Space Agency
ESN	electronic serial number
ETACS	Extended total access communication system
ETSI	European Telecommunications Standards Institute
FACCH	fast associated control channel
FCC	Federal Communications Commission
FCC	forward control channel
FDMA	frequency-division multiple access
FEC	forward error correction

FER	frame erasure rate
FFSK	fast frequency-shift keying
FM	frequency modulation
FMSC	file management subsystem
FOCC	forward control channel
FPLMTS	future public land mobile telecommunications systems
FRC	full-rate channel
FSK	frequency-shift keying
FVC	forward voice channel
G-MCC	gateway mobile communications control center
GLR	gateway location register
GMSK	Gaussian minimum-shift keying
GOS	grade of service
GPS	global positioning system
GSM	global system for mobile communications
HDLC	high-level data link control
HLR	home location register
HMSC	home mobile switching center
HPLMN	home public land-mobile network
HPU	hand-portable unit
HRC	half-rate channel
HUP	handover user part
IBCN	integrated broadband communications network
ID	identification
IDD	international direct dialing
IDO	Nippon Idou Tsushin Corporation
IF	intermediate frequency
IGS	integrated geographical system
IMSI	international mobile subscriber identity
IMTS	improved mobile telephone service
IN	intelligent network
INAP	intelligent-network application part
IPL	initial program load
IQ	in-phase and quadrature channels
ISDN	integrated services digital network
ISI	intersymbol interference
ISO	International Standards Organization
ISUP	ISDN user part
ITA	interim type approval

ITU	International Telecommunications Union
ITW	interworking function
IWF	interworking function
IWP	interim working party
JDC	Japanese digital cellular
J-NAMPS	Japanese narrowband advanced mobile phone system
JTACS	Japanese total access communications systems
LAC	location area code
LAPB	link access procedure balanced
LAPM	link access procedure for modems
LCD	liquid crystal display
LE	local exchange
LEO	low-earth orbit
LMSS	land mobile satellite system
LPC	linear predictive coding
LTP	long-term predictor
MAP	mobile application part
MCC	mobile control center
MCC	mobile country code
MCE	mobile controller equipment
MCN	Micro Cellular Network
MCN	microcellular communications network
MFC	alternative to CCITT system number 7
MDE	modulation and demodulation equipment (modem)
MIN	mobile identification number
MM	mobility management
MMI	man-machine interface
MOC	mobile communications operations center
MONET	Mobile Networks (program)
MOS	mean opinion score
MoU	Memorandum of Understanding
MRI	mobile-reported interference
MS	mobile station
MSA	mobile service area
MSAP	mobile services application part
MSC	mobile switching center
MTBF	mean time between failure
MTP	message transfer part
MTSO	mobile telephone switching office

MTX	mobile telephone exchange
MUP	mobile user part
NA	network architecture
NAM	number assignment module
NAMPS	narrowband advanced mobile phone system
NDC	network destination code
NMC	network management center
NMT	Nordic Mobile Telephone
NRZ	non-return to zero
NSF	nonstandard facilities
NTACS	narrowband total access communication system
NTDMA	narrowband time-division multiple access
NTT	Nippon Telephone and Telegraph Company
NTU	network termination unit
OMC	operations and maintenance center
OSI	open systems interconnection
PA	power amplifier
PABX	private address branch exchange
PAD	packet assembler/disassembler
PBX	private branch exchange
PCH	paging channel
PCM	pulse-code modulation
PCN	personal communications network
PCS	personal communications service
PIN	personal identification number
PLD	programmable logic device
PLMN	public land-mobile network
PMR	private mobile radio
PNA	parallel network architecture
PPR	partial page request
PPS	packet switching service
PSK	phase-shift keying
PSPDN	packet switched public data network
PSS	packet switched service
PSTN	public switched telephone network
PTO	public telephone operator
PTT	public telephone and telegraph operator
QOS	quality of service
QPSK	quadrature phase-shift keying

R/T	radiotelephone
RACE	Research and Development of Advanced Communication Technologies in Europe
RACH	random-access channel
RCC	reverse control channel
RCR	Research and Development Center for Radio Systems (Japan)
RECC	reverse control channel
RELP	residually excited linear predictive (coder)
RES	Radio Equipment and Systems (committee of ETSI)
RF	radio frequency
RFTL	radio frequency test loop
RLP	radio link protocol
RNR	receive-not-ready
ROM	read-only memory
RR	receive-ready
RS	Reed-Solomon (code)
RSS	received signal strength
RT	radio transmission management
RTI	remote terminal interface
RVC	reverse voice channel
RX	receiver
S/N	signal-to-noise ratio
S/No	serial number
SACCH	slow associated control channel
SAK	secret authentication key
SAT	supervisory audio tone
SAW	surface wave filter
SCCH	signaling control channel
SCCP	signaling connection control part
SCM	station class mark
SCP	service control point
SDCCH	slow dedicated control channel
SID	system identification
SIM	subscriber identification module
SINAD	signal-to-noise and distortion
SIS	subscriber identification security
SMG	Special Mobile Group
SMS	service management system
SMS	short-message service
SMT	surface-mount technology
SN	subscriber number

SNAZ	subscriber number allocation zones
SPC	strored program control
SPE	speech processing equipment
SSP	service switching points
ST	signaling tone
STD	standard telephone dialing
TA	traffic areas
TACS	total access communication system
TB	tail bits
TC	transaction capability
TCF	training check flag
TCH	traffic channel
TDMA	time-division multiple access
TG	technical group
TIA	Telecommunications Industry Association
TO	test object
TRX	transceiver
TS	time slot
TSC	transit switching centers
TT	toll ticketing
TTR	test transmitter/receiver
TUP	telephone user part
TX	transmitter
UMTS	universal mobile telecommunications system
UPCH	user packet channel
UPT	universal personal telecommunications
USC	user-specific channel
UTM	universal transverse Mercator
V-MCC	visit mobile communications control center
VAD	voice activity detection
VDU	visual display unit
VLR	visitor location register
VLSI	very-large-scale integration
VMACS	Vodafone Mobile Access Conversion Service
VOX	voice-activated transmitting power switch
VSELP	vector sum excited linear prediction
WARC	World Administrative Radio Conference

Index

The Artech House Telecommunications Library

Vinton G. Cerf, Series Editor

Advances in Computer Communications and Networking, Wesley W. Chu, editor

Advances in Computer Systems Security, Rein Turn, editor

Analysis and Synthesis of Logic Systems, Daniel Mange

A Bibliography of Telecommunications and Socio-Economic Development, Heather E. Hudson

Codes for Error Control and Synchronization, Djimitri Wiggert

Communication Satellites in the Geostationary Orbit, Donald M. Jansky and Michel C. Jeruchim

Communications Directory, Manus Egan, editor

The Complete Guide to Buying a Telephone System, Paul Daubitz

The Corporate Cabling Guide, Mark W. McElroy

Corporate Networks: The Strategic Use of Telecommunications, Thomas Valovic

Current Advances in LANs, MANs, and ISDN, B. G. Kim, editor

Design and Prospects for the ISDN, G. Dicenet

Digital Cellular Radio, George Calhoun

Digital Hardware Testing: Transistor-Level Fault Modeling and Testing, Rochit Rajsuman, editor

Digital Signal Processing, Murat Kunt

Digital Switching Control Architectures, Giuseppe Fantauzzi

Digital Transmission Design and Jitter Analysis, Yoshitaka Takasaki

Distributed Processing Systems, Volume I, Wesley W. Chu, editor

Disaster Recovery Planning for Telecommunications, Leo A. Wrobel

Document Imaging Systems: Technology and Applications, Nathan J. Muller

E-Mail, Stephen A. Caswell

Enterprise Networking: Fractional T1 to SONET, Frame Relay to BISDN, Daniel Minoli

Expert Systems Applications in Integrated Network Management, E. C. Ericson, L. T. Ericson, and D. Minoli, editors

FAX: Digital Facsimile Technology and Applications, Second Edition, Dennis Bodson, Kenneth McConnell, and Richard Schaphorst

Fiber Network Service Survivability, Tsong-Ho Wu

Fiber Optics and CATV Business Strategy, Robert K. Yates et al.

A Guide to Fractional T1, J.E. Trulove

Handbook of Satellite Telecommunications and Broadcasting, L. Ya. Kantor, editor

Implementing X.400 and X.500: The PP and QUIPU Systems, Steve Kille

Inbound Call Centers: Design, Implementation, and Management, Robert A. Gable

Information Superhighways: The Economics of Advanced Public Communication Networks, Bruce Egan

Integrated Broadband Networks, Amit Bhargava

Integrated Services Digital Networks, Anthony M. Rutkowski

International Telecommunications Management, Bruce R. Elbert

International Telecommunication Standards Organizations, Andrew Macpherson

Internetworking LANs: Operation, Design, and Management, Robert Davidson and Nathan Muller

Introduction to Satellite Communication, Bruce R. Elbert

Introduction to T1/T3 Networking, Regis J. (Bud) Bates

Introduction to Telecommunication Electronics, A. Michael Noll

Introduction to Telephones and Telephone Systems, Second Edition, A. Michael Noll

Introduction to X.400, Cemil Betanov

The ITU in a Changing World, George A. Codding, Jr. and Anthony M. Rutkowski

Jitter in Digital Transmission Systems, Patrick R. Trischitta and Eve L. Varma

LAN/WAN Optimization Techniques, Harrell Van Norman

LANs to WANs: Network Management in the 1990s, Nathan J. Muller and Robert P. Davidson

The Law and Regulation of International Space Communication, Harold M. White, Jr. and Rita Lauria White

Long Distance Services: A Buyer's Guide, Daniel D. Briere

Mathematical Methods of Information Transmission, K. Arbenz and J. C. Martin

Measurement of Optical Fibers and Devices, G. Cancellieri and U. Ravaioli

Meteor Burst Communication, Jacob Z. Schanker

Minimum Risk Strategy for Acquiring Communications Equipment and Services, Nathan J. Muller

Mobile Information Systems, John Walker

Narrowband Land-Mobile Radio Networks, Jean-Paul Linnartz

Networking Strategies for Information Technology, Bruce Elbert

Numerical Analysis of Linear Networks and Systems, Hermann Kremer *et al.*

Optimization of Digital Transmission Systems, K. Trondle and Gunter Soder

The PP and QUIPU Implementation of X.400 and X.500, Stephen Kille

Packet Switching Evolution from Narrowband to Broadband ISDN, M. Smouts

Principles of Secure Communication Systems, Second Edition, Don J. Torrieri

Principles of Signals and Systems: Deterministic Signals, B. Picinbono

Private Telecommunication Networks, Bruce Elbert

Radiodetermination Satellite Services and Standards, Martin Rothblatt

Residential Fiber Optic Networks: An Engineering and Economic Analysis, David Reed

Setting Global Telecommunication Standards: The Stakes, The Players, and The Process, Gerd Wallenstein

Signal Processing with Lapped Transforms, Henrique S. Malvar

The Telecommunications Deregulation Sourcebook, Stuart N. Brotman, editor

Television Technology: Fundamentals and Future Prospects, A. Michael Noll

Telecommunications Technology Handbook, Daniel Minoli

Telephone Company and Cable Television Competition, Stuart N. Brotman

Terrestrial Digital Microwave Communciations, Ferdo Ivanek, editor

Transmission Networking: SONET and the SDH, Mike Sexton and Andy Reid

Transmission Performance of Evolving Telecommunications Networks, John Gruber and Godfrey Williams

Troposcatter Radio Links, G. Roda

Virtual Networks: A Buyer's Guide, Daniel D. Briere

Voice Processing, Second Edition, Walt Tetschner

Voice Teletraffic System Engineering, James R. Boucher

Wireless Access and the Local Telephone Network, George Calhoun

For further information on these and other Artech House titles, contact:

Artech House
685 Canton Street
Norwood, MA 01602
(617) 769-9750
Fax:(617) 762-9230
Telex: 951-659

Artech House
6 Buckingham Gate
London SW1E6JP England
+44(0)71 630-0166
+44(0)71 630-0166
Telex-951-659